Get
Me
Out

Get Me Out

A History of Childbirth from the Garden of Eden to the Sperm Bank

Randi Hutter Epstein, M.D.

W. W. Norton & Company

New York · London

For information about permission to reproduce selections from this book,
write to Permissions, W. W. Norton & Company, Inc.,
500 Fifth Avenue, New York, NY 10110

For information about special discounts for bulk purchases,
please contact W. W. Norton Special Sales at specialsales@wwnorton.com
or 800-233-4830

Manufacturing by Courier Westford
Book design by Judith Stagnitto Abbate
Production manager: Anna Oler

Library of Congress Cataloging-in-Publication Data

Epstein, Randi Hutter.
Get me out : a history of childbirth from the Garden of Eden
to the sperm bank / Randi Hutter Epstein.
p. cm.
Includes bibliographical references.
ISBN 978-0-393-06458-2 (hardcover)
1. Childbirth—History. I. Title.
RG651.E67 2010
618.2009—dc22   2009034751

W. W. Norton & Company, Inc.
500 Fifth Avenue, New York, N.Y. 10110
www.wwnorton.com

W. W. Norton & Company Ltd.
Castle House, 75/76 Wells Street, London W1T 3QT

1 2 3 4 5 6 7 8 9 0

To Stuart, Jack, Martha, Joseph, Eliza

Contents

Part Three

Part Four

Part Five

Introduction

In 1889, a woman in northern England gave birth to a boy whose head came out draped inside a piece of the amniotic sac. Typically, if a piece of the sac sticks to the newborn's head, a nurse washes it away. Long ago, they were called cauls and were considered either really lucky or really ominous, depending on your birth guru. Some people carried their cauls for good luck because they were supposed to prevent drowning. You could buy someone else's if you weren't lucky enough to be born with one.

There was a market in amniomancy—forecasting the baby's life by "reading" his caul—and it goes back a long way. In AD 208, a Roman emperor's son was born with a sac on his head, and everyone celebrated because they figured that the boy was destined for greatness and by extension so was the family's dynasty. As predicted, the boy joined his father as co-emperor by age 9, but father and son were murdered the following year. Whatever. Caul believers continued to keep the faith. By the Middle Ages, doctors poked fun of the folksy caul forecasters, but their cynicism did not deter those who wanted to believe, who wanted to see the future in pregnancy leftovers. In 1658, Sir John Offley bequeathed his daughter "a Caul that covered my face and shoulders when I first came into the world."

As it happened on that day in 1889, the local newspaper reported that "the greatest excitement prevailed in the neighborhood, some of the women declaring that nothing short of a miracle had been enacted." Not only was the baby's head stuck inside his sac, but according to the local newspaper, the caul had these words inscribed on it: "British and Foreign Bible Society." The writing was on the wall, or the sac, as the case may have been. Except for one glitch. A skeptical doctor investigated the labor room and realized that the caul had been placed on a Bible with those very same words embossed on the cover. No matter, the townsfolk still considered it propitious. No one let the facts get in the way of a miracle. Or, as the *Leeds Mercury* reported, "The inhabitants still ascribe the affair to supernatural influence and declare that the child is a 'missionary born' and that they will evidentally watch his career with a great amount of interest."

Thomas Forbes, the late Yale anatomist and medical historian, once said that the history of obstetrics is in large part a history of superstition. He had a point. In ancient times, men who wanted a boy drank red wine spiked with dried rabbits' wombs and read guidebooks that told them how to have sex to ensure simultaneous orgasms, considered a must for conception. Pregnant women tried to think happy thoughts to make happy babies and avoid staring at ugly things to prevent ugly babies. As Forbes explained, in the old days "before scientific explanations were available, speculation had to suffice, and superstition was often a result."

True enough. But how far have we really come? I don't mean to suggest that every ultrasound and piece of advice is based on superstition. Childbirth, then and now, is a wonderful blend of custom and science. *Get Me Out* is about the evolution of advice and the advances; the choices couples have been offered; and the decisions they have made. The way expectant parents have chosen to give birth—or the way they have negotiated with their caregivers—reveals the lengths they will go in their desire, at times desperation, to create the ideal offspring. This is a book about the history of the fears, disappointments, tragedies, and glees. It is about birth, which includes sex, life, and sometimes death. *Get Me Out*, then, is not a catalog of old-fashioned and often strange words of wisdom, but an examination of how ideas about pregnancy and birth shed light, directly and obliquely, on contemporary society.

Sometimes birth choices are not about perfection or fears but just the way we happen to be living then. In Colonial days, when a woman's job was to get married, make babies, and in between pregnancies help her parturient friends, new mothers stayed in bed for weeks on end. They called the postpartum phase "confinement," because women were confined. Toward the end of the twentieth century, when women wanted to get back to work (or at least out of the labor room), it became medically okay to get out of bed right away. Were there scientific studies to prompt the change? I should think not. We did what made sense for our lifestyle.

Sometimes it's about how we imagine the ideal birth. There was a time in the early 1900s when feminists demanded their right to be knocked out with drugs during delivery. They and physician advocates claimed that the drugs were healthier for mom and for baby. Contrast this to the late 1900s, when feminists railed against painkillers that made them miss part of the birth experience. And they, too, claimed that their way of a drug-free birth was healthier for mom and for baby.

Sometimes it's a statement about how couples think pregnancy should be experienced. When Elisabeth Bing first introduced Lamaze to America in the early 1950s, the few folks who heard about her crusade thought it was a silly idea and even fewer tried it. Bing said that the men and women who attended her classes in the early days went because they figured it couldn't hurt, not because they believed. By the 1970s, she was teaching to a sold-out celebrity-studded audience who had faith in the system. Nowadays, Lamaze is a household name, but few husbands are counting in the Lamaze way by the bedside, and much to Bing's chagrin, they do not have time for weekly seminars. There was a time in the late 1990s when she was flooded with calls from harried couples who wanted to learn her tricks, but the CliffsNotes version, a quickie one-night session.

Get Me Out begins with Eve, because she started the whole birth-is-painful thing, and concludes with young couples today who sperm-shop and freeze eggs for one reason or another. While I include a few chapters on birth in the early years, the crux of the book focuses on the late nineteenth and the twentieth centuries in America because that's when everything started to change. Scientists discovered germs, and shortly thereafter doctors decided birth should be hygienic. Feminists crusaded for women's rights and began telling birth attendants what to do. Obstetricians went from placing a stethoscope on the belly to listen for the pitter-patter of a heartbeat to using ultrasonography to snap a 3-D image of the fetal heart. Birth went from home to hospital, from drug-free to drugs on delivery, from midwives to doctors, from the occasional C-section to C-sections on demand. On the most superficial level, the huge changes in the birth process reflect the rise of the urbanization, feminization, and technologicalization of American society. On a more profound level, the way we give birth is a story about our deepest desires and our fundamental concerns about life, death, and sex. We are creating life and we will do anything that we think is in our power to control and to perfect the outcome.

Get Me Out is not an advice book. It won't tell women what to eat or which test to take, and it won't tell men how to coach from the sidelines, but the stories and tragedies of yesteryear will prompt readers to be more inquisitive about health decisions today. The guidance that I hope you glean from the book should pique your curiosity to think about the medical maze in a different sort of way, to ask deeper questions, and to question yourself about the choices you make. Is it because of a new study or a new medical fashion?

A familiarity with the bizarre history of pregnancy is sure to infuse a healthy skepticism among patients as they navigate the murky, often dubious medical advice. The lessons learned have far-reaching implications not confined to the nine months of gestation. A few themes percolate throughout: I argue that the quest for the perfect offspring was a goal since antiquity, not simply a notion prompted by modern medicine. I also illustrate the ways in which we have always believed in the intricate ties between emotions and the physical body.

The aim of these seemingly diverse stories is to portray the field of gynecology as a marriage of medicine and society. It is an arranged unit at times maturing into a meaningful relationship, at times fractious. Yet the intimate and inseparable ties of science and society always spawn new notions of women's reproductive health and sometimes produce life-preserving and lifesaving technology. As Judith Leavitt, a professor of women's studies and medical history at the University of Wisconsin, put it, "Childbirth symbolizes the historical and cultural definition of a woman's essence."

What makes pregnancy different from other medical investigations is that we are, for the most part, dealing with healthy women. Pregnancy is not a disease, and yet, we do at times need medical assistance. Its history bares clues about how women and their doctors think about health and disease. How a person is labeled (sick versus healthy) is not simply the result of a medical test that can detect lurking germs or cells gone awry. It has just as much to do with the way people think about wellness and what comprises a healthy body. When did pregnancy, once considered just an ordinary part of a woman's life, become pathological, a diagnosis requiring special expert-supervised therapy? Prenatal care is certainly a modern concept. What spurred that idea? All of our health decisions are far from knee-jerk reactions to medical advances, but responses to political forces, habits, current mores that may or may not be to our benefit.

And while it may sound trite, I chose to explore the history of childbirth because I am truly mystified by the science of it all. A burst of human life from two half sets of genetic material is nothing short of a miracle. (One of my favorite reporting days was in a fertility clinic, where I watched millions of sperm swimming toward an egg. The one gargantuan egg seemed like the all-mighty Oz chased by zillions of tiny munchkin sperm.) I find it striking that for all of the modern technology available today, for all of the ultraclear pictures we have, we are still in the dark about so many things that go into making babies. No one knows precisely what triggers those very first baby-making cells to merge and transform from microscopic mass to person. Sure, we've discovered the receptor on the ova that attaches to the sperm, and we've seen the changes that take place to prevent another sperm from cracking into the egg. And yet, there is still so much more to discover. Why this sperm and that egg? Why did this bundle of cells survive the journey from fallopian tube to womb to world and the others not? Ever since the very first zygote snuggled into the wall of the uterus, transformed into an embryo, then fetus, and ultimately a baby pushing into the outside world, scientists, philosophers, and doctors have created theories to explain the process from beginning to end. Their theories have dictated the way couples have tried to have babies and the way women behave once they know they are pregnant. And we still have so much to learn.

We've come a long way since the days of caul forecasting. And yet, despite all the changes, all the high-tech stuff, we are still the same worried parents as our great-great-great-grandparents, wondering what we can do to tinker with Mother Nature to up the odds of having a healthy baby or the next gold medalist. We still like to believe that somehow we can control the situation by food, thoughts, or drugs. And we still blame ourselves when things go awry. We still like to share our birth stories, hoping to glean information or spread the gospel. Linked together through the ages, these stories become a chronicle of our past that enlightens our future.

Part One

1

Eve's Doing

Birth from Antiquity through the Middle Ages

E ve, the first woman to become pregnant, suffered from excruciating pain during the delivery because she cheated on her diet. God told her to not eat an apple, but she was tempted by the serpent's claim that the forbidden fruit would endow her and Adam with worldly knowledge. In God's fury, he transformed the serpent into a belly-crawling creature. Then he turned to Eve and said, "I greatly multiply your pain in childbearing; in pain you shall bring forth children."

The thought pattern was set. Women deserved pain. In 1591, Eufame Maclayne was burned at the stake for asking for pain relief during the birth of her twins. Attitudes did not change much when safer anesthetics were discovered in the middle of the nineteenth century. Most people thought they were fine for surgery but not childbirth. Devout men and women believed that the pain in childbirth was a heavenly duty. If you couldn't endure the agony of childbirth, how would you handle the ups and downs of motherhood? (Why no equivalent hazing process for fathers? Vasectomies without pain meds?) Pain relief became somewhat

acceptable when Queen Victoria asked Dr. John Snow for a whiff
of chloroform to ease her delivery during the birth of Prince Leo-
pold on April 7, 1853. But only somewhat.*

Eve, of course, had a lot more to think about than labor
pains. She was the only woman in the history of the planet to go
through pregnancy without any advice, solicited or otherwise. We
don't know whether Adam was nagging her to eat certain things
or avoid others, but given how easily she manipulated him into
eating an apple, it doesn't seem like he was the one wearing the
pants in the relationship. Eve had no one. No mother. No guidebook. No friends with their own birth stories. Instead, she suffered the punishment. Despite the dire consequences—having to
squeeze babies through an impossibly teeny orifice—she popu-
lated the earth, launching one of the greatest traditions of wom-
anhood: feminine determination. She got to have her apple and
her babies, too.

As soon as her daughters and her daughters' daughters reached
childbearing age, none of them would ever experience pregnancy

---

* Steven Johnson's *The Ghost Map*, a book about cholera, includes Snow's
account of the ether queen, gleaned from his letters at the Wellcome Institute
for the History of Medicine: "Thursday April 7: Administered Chloroform to
the Queen in her confinement. Slight pains experienced since Sunday . . . I commenced to give a little chloroform with each pain, by pouring about 15 minims
[0.9 milliliter] by measure on a folded handkerchief. The first stage of labour was
nearly over when the chloroform was commenced. Her majesty expressed great
relief from the application, the pains being very trifling during the uterine contractions, and whilst between the periods of contraction there was complete ease.
The effect of the chloroform was not at any time carried to the extent of quite
removing consciousness. Dr. Locock thought that the chloroform prolonged the
intervals between the pains, and retarded the labour somewhat. The infant was
born at 13 minutes past one by the clock in the room (which was 3 minutes before
the right time): consequently the chloroform was inhaled for 53 minutes. The placenta was expelled in a very few minutes, and the Queen appeared very cheerful
and well, expressing herself much gratified with the effect of the chloroform."

without a bombardment of words of wisdom. We seek them. They seek us. Who did Eve's children turn to? Our ancestors did what women have been doing all along. They turned to each other and self-proclaimed birthing gurus. They turned to medical men, the presumed pillars of knowledge. The literate few could read guidebooks—rather, guide papyri. You may think life was easier for our great-great-great-grandmothers, given the narrow range of advice. But from their perspective, it was a dizzying whirligig of do-this-don't-do-that.

Birth from antiquity through the Middle Ages was an all-girls affair orchestrated by men who had never seen a baby born. It was considered obscene for a man to enter the delivery room, yet they wrote the guidebooks, doling out advice based on hunches handed down over generations. (In 1522, Dr. Wert, a German doctor, was sentenced to death when he was caught dressing like a woman and sneaking into a delivery room.) Their words of wisdom (or of ignorance) were a man-made concoction of myth, herbs, astrology, and superstition. Nearly everything was about good sex and good thoughts and eating and drinking the right things. It was not simple. As far back as 1500 BC, probably even earlier, women had access to all sorts of explicit information about sex, pregnancy tests, abortions, and contraceptives.

What women went through back then, the whole experience, must have been one big guilt trip. Should anything have gone wrong, there were so many reasons to blame your own behavior. Did I do something that deserved God's curse? Was I drinking too much wine? Did I harbor evil thoughts?

If you were lucky to be in a city, you may have been helped by a licensed midwife (European cities started educating and registering midwives around the fifteenth century); if you were in the rural outback, you may have had an uneducated but experienced midwife or a female family friend. In any event, you were surrounded by a gaggle of women. Oddly enough, expectant women were not supposed to be catered to, but to cater. You were expected to act as hostess and serve the aptly coined "groaning beer" and "groaning cakes." Friends of the laboring woman were called "gossips," as in God sibs, as in siblings of God. You can assume they did what all women would do under the circumstances—sit around and talk about other people. So what was once an epithet for "close-to-God" morphed into a term for "behind-the-back chatter."

For the first millennia or so, women relied on the same traditions written and rewritten, told and retold, with very little change. Centuries after Hippocrates, Aristotle, and Galen were long dead, doctors were rewriting their words of ancient wisdom with little thought to the fact that the wisdom may be outdated. Medical authors were scribes, not enlightened experts, and certainly not investigators. Pregnancy advice in antiquity was virtually the same as the advice doled out generations later to medieval women. Sometimes experienced midwives learned a thing or two to tweak the process, but the books did not change.

Women were told how to speed labor (a concoction of herbs), what to eat (nothing too spicy), what to drink (not too much wine), and what to think (no angry thoughts). Women were told how long to breast-feed and when to hand the baby to a wet nurse. They were told to have enough sex because a splash of sperm moistens the womb. They were also told not to have too much sex because it wears out the baby-making machinery. That's why "whores have so seldome children," one guide said, because "satiety gluts that womb." In France, pregnant women rarely left the house after dark because they were told that if they looked at the moon, the baby would become a lunatic or sleepwalker.

One guidebook prescribed his and hers cocktails to up the odds of having a boy: red wine tainted with pulverized rabbit's womb for him; red wine with desiccated rabbit's testicles for her. Were couples truly doing shots of this stuff? We'll never know. But couples who wanted children—or preferred one sex to another—were willing to try anything. Think about the hormones we're shooting ourselves up with today. Maybe dried testicles wasn't so weird after all.

France's sixteenth century queen Catherine de Medici had the money and wherewithal to get all kinds of medical advice and treatment when she could not get pregnant. She chose first her folk healer, who told her to drink mare's urine and to soak her "source of life" (vagina?) in a sack of cow manure mixed with ground stag's antlers. The king was never sexually attracted to his wife. The dung diaper could not have helped the situation. Eventually, de Medici went to a doctor who diagnosed the teenage royals with physically defective reproductive organs. He had a different cure in mind. No one knows what it was, but it worked. They went on to have nine children. In between the folk and medical wisdom, she tried her own tactic: she commanded her servants to drill a hole in a floor so she could watch her husband have sex with his mistress and learn a thing or two about baby making. Maybe that did the trick.

Despite the collection of advice manuals, we really have no idea whether our great-great-great-grandmothers followed advice doled out by mothers, midwives, or medical men. Did they second-guess their doctors when their mothers told them something else? Did they read the pregnancy books or just let them collect dust on a shelf? If Eve is supposed to be a female representative, a role model of sorts, then perhaps there is something innately feminine to questioning authority. You have to assume that we have always balanced experts' suggestions with the advice given by friends and mothers, not to mention our own gut instincts. What was written in the books, then, may not tell the story of birth in antiquity, but it expresses what was important to women and caregivers about the birthing process.

The great thing about consistent medical knowledge is that doctors could reprint their books forever. Today's medical textbooks are outdated by the time they go to press. Soranus was a famous Greek physician who wrote the definitive book on gynecology in the second century. It was *the* leading text for the next thousand years. Not a bad run. The first part was matchmaking advice. He told men how to choose a fertile partner. Here's what to look for: a cheerful woman who is not mannish or flabby; a woman who digests food easily; a woman who does not have chronic diarrhea. Constipation, according to the book, suffocated the fetus. Diarrhea washed it away. Or, as he put it, women with chronic bowel issues would never be able to "lay hold of the seed injected into them." It's a wonder how men approached the subject of bowel movements on the first date. Soranus also told men to date a normal woman with a normal uterus. No further advice about how you figured out who had the normal uterus or what constituted a normal uterus in the first place—or, for that matter, what constituted a normal woman.

> Sex during pregnancy produced kids who were "ill tempered, sickly, and short-lived."

Soranus thought moderate drinking a good thing, but too much was dangerous. He believed that your thoughts molded the growing baby, so if you got drunk and had oddball fantasies, you would have weird children. There was proof. He said one women thought about monkeys during a drunken sexual escapade and her kids turned out hairy. On the other hand, an ugly man from Cyprus made his plain-Jane spouse stare at a beautiful statue during sex. Wouldn't you know it? They had gorgeous children.

Sex during pregnancy was considered dangerous to the growing fetus because it drained a woman's vital juices that should flow to the baby. Too much intercourse caused children who would be "defective in vital and other qualities, ill tempered, sickly, and short-lived." Smart parents made smart children, but again, only if they did not have too much sex. Otherwise their little ones, though born with higher than normal intelligence, would be weaklings and die before the age of 10. As always, moderation was key, but no one said what was the normal amount of sex.

Soranus wrote about positions and maneuvers to up the odds of conception. He also gave birth control advice, that is, how to have sex without conceiving: After your partner ejaculates inside of you, hold your breath, forcefully sneeze and then drink ice-cold water. If that failed, he recommended Hippocrates' abortion remedy: kick your heels into your buttocks until the seed drops out.

He also promoted do-it-yourself home pregnancy tests. Many tests relied on urine chemicals, much like today. He told women to urinate on a bouquet of wheat, barley, dates, and sand. If the grains sprouted, you were pregnant. If wheat grew, it was a boy; if barley, a girl. The sex selection may have been nonsense—maybe not—but perhaps the surge of the pregnancy hormones fertilized the sprouts.

Many of the earliest women's health books were written by monks, the very people who had the least use for the information. One book instructed men to get women in the mood by rubbing her between vulva and anus. What good this did monks is anybody's guess. One of the most popular monk guides, *Women's Secrets*, or *De Secretis Mulierum*, has been translated from the original text into modern language by Helen Rodnite Lemay, a medieval scholar. As she sees it, the book not only doled out health advice, but showed

The vagina is the "antechamber to lodge a Man's Yard."

what some of the great thinkers, including Hippocrates, Aristotle, Soranus, and Galen, thought about women. In a nutshell, not much. She also showed that while health care was a combination of philosophy and medicine, philosophers and doctors perceived the human body and the cures in vastly different ways.

An eleventh-century birth manifesto was supposedly written by a midwife named Trotula. To this day, no one knows whether the book represents a collection of writers, whether there really was such a person, or whether the author was a man or a woman. It, too, was filled with prescriptions for herbs and wine concoctions.

Aristotle, the philosopher, believed a man's "seed" shaped menstrual blood into a human being. Doctors reconsidered. They thought that during pregnancy the menstrual blood flowed upward and turned into breast milk. They also thought that men provided the life source that created the human. Women were mere baby-making vessels. Or as a sixteenth-century expert put it, the purpose of a vagina was to be an "antechamber to lodge a Man's Yard." A penis was called a yard, as in a 36-inch stick. Definitely wishful thinking. Women were either hollow or had a few baby-making parts within. Such thinking explained Hippocrates' odd fertility test: Put a scented cloth over the vagina. If the same odors wafted out the woman's nose, she was barren because there was nothing inside to block the aroma. Doctors also knew from early on that women were only fertile certain times of the month. But they had it mixed up. They thought women conceived during menstruation.

The second-century doctor-philosopher Galen is revered for a number of reasons, including the fact that he figured out that blood travels in arteries. Impressive stuff. But our female ancestors have a much better reason to thank the Great Thinker. He also promoted the idea that orgasms were essential for conception. So what if he was wrong. Was it really such a bad thing? This

underlying message—good sex made for good babies—turned
ancient baby-making manuals into sex guides. His ideas, errone-
ous or not, would flourish. Think soft porn—not a what-to-expect
papyri. (Getting some things closer to the
truth, he claimed that both men and women   "It often hap-
produced seeds, which was contrary to the   pens that a
accepted dogma that the man pumped his   woman con-
life force into the hollow woman.)   ceives if she

Generations before the Puritans made   is in a bath
sex talk nasty; centuries before America's   where a man
own nineteenth-century Anthony Com-   has ejaculated
stock enacted laws making even the word   because the
contraception the equivalent of pornogra-   vulva strongly
phy, newlyweds—and single women for   attracts the
that matter—could enjoy reading detailed   sperm . . . This
descriptions about sex, all in the name of   has been
family planning. Galen warned men that   attested to by
they had better satisfy their mate or else   experience."
she'd never get pregnant and she would
probably seek satisfaction elsewhere.

Reading one fifteenth-century birthing
guide makes you think of an erotic game of twister: Head low,
hips high, left foot tucked under the hip, right foot extended, and
have an orgasm at the same time as your partner. Was this done in
the privacy of your own bedroom, or with a spinner and a bunch
of friends?

Jane Sharp, a British midwife writing in 1671, said, like so many
others had before and after, that penis size affected fertility. Too
big was just as bad as too small. A penis longer than 11 inches
would spray the womb with seed, somehow mucking up conception. A penis shorter than 1 inch would not touch the opening of
the womb, so the female and male seeds would never mix. Within

the normal penis range, between 1 and 11 inches, the vagina acted like spandex, stretching and shrinking to suit the man—or as the experts put it, the vagina would "dilate, contract, extend or abbreviate itself according as it is necessary to bear exact proportion with the bigness or length of the man's yard . . . and by that extent the pleasure may be mutually augmented." Because the clitoris was considered nothing more than a female version of a penis, doctors worried about clitoral size too—not for fertility, but because a woman with an oversized clitoris was destined to lesbianism. British midwife Jane Sharp claimed that there was no such things as British lesbians; it was something that happened to large-clitoris women in the Indies and Egypt.

There were remedies for small-penis men. Sharp told them to eat beans and roughage because "windy spirits" inflated the penis. Another doctor suggested a do-it-yourself treatment of horseradish cream, rubbed three times a day for 40 days. If you did not want to do it yourself, you could ask your doctor to do it for you. Talk about doctor-patient relationships.

Babies were born from a woman's "voluptuous itch" to copulate. Why else would any woman dare experience life-threatening pregnancy unless she just couldn't help herself when passions overwhelmed her? One guidebook told women that getting pregnant was the same as catching a serious disease—an opportunity to die. To make matters worse, it told readers that motherhood was a downhill slope into ugliness and old age, "a loss of beauty, which is the most precious gift she has."

When a woman wanted sex, her womb would open and allow the male seed to enter, facilitating baby making. In the words of a sixteenth-century guide, "When they have an appetite the womb desirous and covetous the seed, at that instant opens to receive and be delighted with it." Pleasure is due to four "carruncles" (beads? bumps?) lining the opening of the womb that "close more pleasantly upon the Man's Yard, whereby the Woman is also more delighted." Likewise, if you're not having a good time, your womb stays shut.

The womb seemed homeless, wandering all over the body, bobbing up and down like a yo-yo. A "suffocated" womb could not maintain a pregnancy. Alas, there were cures. You could sniff awful-smelling fumes to scare the uterus back to its rightful position, as if the uterus had a mind of its own and would run away from the stench. (Given that animal dung and menstrual blood were used as standard therapies, the treatment must have been utterly repulsive.) Another option, for married women only, was for the husband to "possess her roughly, taking care that she assumes the bottom position, and bring on orgasm in this fashion." The vaginal secretions would wash away noxious fluids. The third and most intriguing option was a visit to an obstetrix, a doctor's assistant who brought women to orgasm. One wonders how they advertised the job.

The vagina, too, was endowed with powers, able to lure sperm even without penetration. That explained how virgins became pregnant—or it gave pregnant teenagers the best excuse ever. A girl could get pregnant bathing in a tub where a man had recently ejaculated. The vagina sucked up sperm like a vacuum. Those monks who wrote *Woman's Secrets* put it this way: "It often happens that a woman conceives if she is in a bath where a man has ejaculated because the vulva strongly attracts the sperm. . . . This has been attested to by experience." Along the same lines, it was said that if a cat ejaculated on sage and then a man ate the sperm-tainted herb, he would grow a cat in his stomach and vomit it out. Sounds more Harry Potter than birth manual.

The blockbuster pregnancy guide of the Middle Ages was written by Dr. Eucharius Rösslin. He must be considered one of the greatest marketing wizards ever. Rösslin was a government doc-

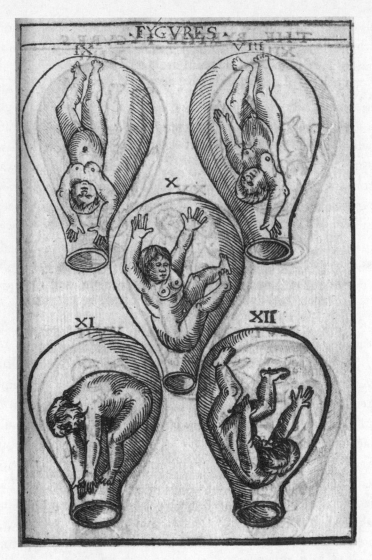

Sixteenth-century illustration of fetuses in the womb. Eucharius Rösslin,
*The birth of man-kinde; otherwise named The womans booke. Set forth in English by Thomas
Raynald physitian.* London: Printed for A. Hand and to be sold by Iames [James]
Boler, 1626. Courtesy of the Malloch Rare Book Room, New York Academy of
Medicine Library.

tor in Germany responsible for tracking epidemics and licensing midwives. He was a statistician. And yet, his greatest insight was recognizing the power of the printing press. Others may have dismissed the gadget, but Rösslin suspected that a book about birth would be a surefire hit, particularly if it were written in easy-to-digest prose. Most *The Rose* medical books were written in Latin and *Garden for* aimed at an expert audience, not for patient *Pregnant* consumption.   *Women and*

He was right. No matter that he never saw *Midwives* a baby born, studied childbirth, or was even up-to-date on contemporary practices. *The Rose Garden for Pregnant Women and Midwives*, the first book focusing solely on pregnancy and childbirth, was published in 1513. It was a best seller for 200 years, translated into at least five languages.

He claimed to have called his book *The Rose Garden* because a woman's fertility is her own special rose garden. He may have named the book after himself. Rösslin means "little rose." The English edition, translated by Thomas Raynalde, was called *Byrth of Mankynde*. The book is divided into 12 sections, including chapters about how babies look in the womb (mini-adults in various acrobatic poses); how a woman should behave before, during, and after delivery (kindly to everyone around her); how a woman could tell if she was going to miscarry; and how long a woman should nurse her baby.

The only problem—for women, not for sales—was that he printed obsolete information. *Rose Garden* dished out the old stuff harking back to AD 100, techniques no longer used by many sixteenth-century midwives. To remove a stillborn child, he told women to shove a dough made of dung, cow bile, and 25 herbs into the vagina. Even the illustrations are copies of ninth-century drawings. To hear medieval scholar Wendy Arons describe it, the only thing Rösslin's book accomplished was marking "the beginning of the encroachment of the midwives' domain by the male medical professional." Doctors may not have changed their notions of childbirth in the past millennia, but midwives were beginning to drop some fashions in favor of others. Arons has a hunch that Rösslin may have written his book solely to reinforce male expertise. He penned a poem in his best seller that was certainly more about midwife trashing than birth guidance:

> *I'm talking about the midwives all*
> *Whose heads are empty as a hall*
> *And through their dreadful negligence*
> *Cause babies' death devoid of sense.*
> *So thus we see far and about*
> *Official murder, there's no doubt.*

Suffice it to say, by the end of the 1500s, doctors were beginning to bad-mouth female helpers as a way to make room for themselves in this females-only business. Still, the birthing enterprise would not change dramatically until the advent of obstetric tools. These tools gave men something concrete that they had and that midwives lacked, and it would help convert birth from what was so long a spiritual journey to a medical procedure.

2

Men with Tools

Forceps Use from 1600s to 1880s

any families have secrets. Most of them have to do with the crazy uncle or the embarrassing cousin. The Chamberlen family guarded a medical secret—and not the kind that has to do with weird defects passed from generation to generation. For 200 years, beginning in the late 1500s, this family of doctors concealed a design for obstetric forceps.

During the Chamberlen years, talk was that these men had the highest success rates for getting the baby out safely and keeping the new mother alive, thanks to their secret tool. Despite the buzz, no one ever saw their instrument, not even the doctors' closest confidants. This family did not share their medical know-how with anyone. If they went public, they'd lose the Chamberlen advantage, one that landed them steady work with royalty and earned them phenomenal wealth.

The Chamberlens were sneaky doctors who vied for wealthy clientele. Women loved them because they believed that a Chamberlen in the birthing room ensured a safe delivery. Colleagues despised them because they saw the Chamberlens as arrogant and conniving. One critic wrote an anti-Chamberlen poem that concluded:

> *"To give you his character truly compleat*
> *He's Doctor, Projector, Man-Midwife and Cheat."*

The Chamberlen story is littered with rumor and hyperbole, but even if we dismiss the juicier bits, the legend of the forceps family reveals how doctors used birthing instruments to convert childbirth from a natural state of affairs to a medical event. And it shows how doctors, beginning in the late sixteenth century, created the specialty of obstetrics.

Pre-forceps birth was a social and spiritual event. The midwife—the female midwife—crouched in front of the laboring woman rubbing the perineum with oils and herbs. Often, she used a birthing stool, a doughnut-shaped chair that allowed gravity to help the baby slip out. No men allowed. Post-forceps, men would gradually become a routine presence. Doctors did not like stools. They preferred the patient lying in bed where they could use their tools with ease. Of course, we cannot credit the forceps with changing everything. But the forceps certainly hastened the change and opened the door for other birthing gadgets.

What the Chamberlens gave the world, as opposed to the devices made earlier, was a gentler sort of instrument, one trumpeted to get the baby out in one piece—and alive. One Chamberlen commented that forceps dispelled the old adage that "when a man comes, one or both must necessarily die." He meant that before forceps, men entered the birthing room only if the parturient woman, the baby, or both were near death.

Were women relieved when medical men came to help? That's up for grabs. On the one hand, they were happy to have someone who came with alleged expertise. On the other hand, they knew that doctors came with a medical bag of spooky tools. Either way, for the first time, women had to think about how private they wanted the birth of their babies. When do you call for extra measures and when do you let nature take its course? It's a debate that began in the late sixteenth century but resonates loudly and clearly into the twenty-first.

Before forceps, babies stuck in the birth canal were dragged out by the doctor, often in pieces. Sometimes midwives cracked the skull, killing the newborn but sparing the mother. Sometimes doctors broke the pubic bone, which often killed the mother but spared the baby. Doctors had an entire armamentarium of gruesome gadgets to hook, stab, and rip apart a hard-to-deliver baby. Many of these gadgets had an uncanny *"the woman is to be encouraged with hope by kind language"* resemblance to medieval torture tools. A few looked like fireplace stokers, and one looked like a gigantic cast-iron corkscrew.

Ancient brochures provided explicit instructions on how to use knives and hooks to carve up the dead baby to make it easier to extract. In the 1300s, the Japanese used a contraption made of nets and whale bone to hook a baby lodged in the vagina. The French had a poker-like instrument that stabbed a hole in the baby's head and dragged it out. Speed was of the essence because some doctors believed that dead fetuses exuded noxious fumes that poisoned the mother. Babies who did not come out easily lacked a vital spirit, but they were not deprived a spiritual death. Catholic women could request an in utero baptism, often performed by a midwife, who squirted holy water into the birth canal premutilation. An Ayurvedic text told midwives that "the woman is to be encouraged with hope by kind language"—that is, before the midwife rips the baby to shreds.

Even the word *forceps* comes from something that sounds more like a kitchen appliance than a birthing tool. It comes from the Latin words *formus*, meaning "hot," and *capere*, "to seize." You were grabbing hot things—like a chicken off the grill.

The Chamberlens' British-based baby business began in 1569, when Dr. William Chamberlen sailed to England with his pregnant wife and three children.* It was a grand idea for the Chamberlens to get out of France ASAP. They came not for the English countryside, but because they were Huguenots and no longer welcome at home. Continental Europe would be in the throes of war for the next hundred years. Peaceful England beckoned.

"These grasping tightwads bent only on their own enrichment,"

To this day, historians haven't figured out which Chamberlen among a gaggle of man-midwives was the one to design the forceps. If you had to guess, your best bet would be Peter Chamberlen. That's because there were so many of them. Nearly every other Chamberlen was named Peter, making life miserable for historians trying to figure out which Peter did what. Dr. William, for one, had five children. He named two of them Peter and both became doctors. It's thought that one of them, probably the older one, designed the forceps.

The original Peters were not even full-fledged physicians, but members of the Guild of Barbers and Surgeons, a union of sorts that included all kinds of cutters, whether they styled hair,

---

* Chamberlen does not sound French. The spelling probably evolved over the years. One early document refers to the earliest immigrant as Dr. Chamberlaine, another as Chamberlayne. K. Das, *Obstetric Forceps: Its History and Evolution* (St. Louis: Mosby, 1929). For further reading about Huguenot history—and what life must have been like for the Chamberlens before they left France, see Mack P. Holt's *The French Wars of Religion, 1562–1629*, 2nd ed. (New York: Cambridge University Press, 2005), which relates a lively easy-to-read story of the French civil wars. Leonie Frieda's *Catherine de Medici* (New York: Fourth Estate: HarperCollins, 2005) is a thorough and sympathetic biography of the Italian-born French ruler.

trimmed beards, or removed gallbladders. You could do all three. Some barbers were known to cut hair and bloodlet on the side. The Peters got into trouble with the medical profession again and again. They rarely attended the mandatory meetings of the Barbers and Surgeons Guild. Worse, they often gave out drugs without the mandatory okay from the Royal College of Physicians. Lowly surgeons, considered more butcher than healer, had to get permission from *real* doctors before they prescribed medicine. (To this day, you can hear remnants of the class distinction among doctors in England. Surgeons are addressed as "Mr." and all other physicians "Dr." (Because of the recent reversal of status, a doctor called "Mr." garners just as much, if not more, prestige as a doctor called "Dr.") During the days of first-generation Chamberlens, surgeons did the bloody stuff that elite doctors avoided. The older brother flouted the drug rule so often that he was thrown in jail in 1612. According to Chamberlen lore, either the queen got him out of jail and presented him with a diamond ring or he got out thanks to the Lord Mayor of London and the Archbishop of Canterbury. Either way, he got out. Despite their claims of expertise, the Peters would never have been admitted to the prestigious royal society. They lacked key credentials—a degree from either Oxford or Cambridge.

When the younger Peter got married and had children, he named his son Peter. This young Peter collected degrees from Padua, Oxford, and Cambridge universities within three years and also became a member of the elite Royal College of Physicians— that is, after he was rejected twice. First time around, they said he was too young. Second time around, they balked at his inappropriate attire. They told him to ditch the "gaudy, frilly" look and go for appropriate adult clothes. Years later, he would be kicked out.

With credentials from leading medical schools plus Royal College membership (albeit temporary), Peter was eyed as the perfect

relative to take over his Uncle Peter's lucrative midwifery practice that catered to aristocracy. Uncle Peter never bothered with
poor people and has been accused of milking the rich. One writer
called the Chamberlens "grasping tightwads, bent only on their
own enrichment."

Armed with their tool, touted to speed deliveries and save babies,
the Chamberlens became known as the best "man-midwives"
around. Man-midwife is what men decided to call themselves.
These self-proclaimed experts had varying degrees of degrees.
Some man-midwives had no formal education at all. Some were
graduates of Oxford or Cambridge, but that did not ensure they
ever learned anything about childbirth.*

Unlike female midwives, who were coaxing and massaging,
men did the job practically blindfolded in accordance with etiquette of the day. Some men considered what the man-midwife
did tantamount to adultery. To appease worried husbands, the
doctor covered the parturient woman with a huge sheet, one end
wrapped around the mother, the other tied around his neck. Doctor and mother were the posts holding down a big tent. Groping
around without looking, the doctor was supposed to get the baby
out. Add to this the burden of using forceps. The doctor had
to manipulate his instrument into the right openings and nudge
a baby out, all without stabbing the fragile newborn or ripping
mom. Needless to say, there were a lot of accidents. To make matters worse, many doctors had no idea what they were doing under

---

* Women who delivered babies, in contrast to men who delivered babies, were
at risk of being accused of witchcraft should anything go wrong. It's been said that
many of the witch trials were instigated by men trying to take over the baby business. Ann Hutchinson, who would eventually have a highway in New York named
after her, was deemed a witch when she helped a friend give birth and the baby
died. According to the *Washington Post*, Grace Sherwood, another midwife-witch,
was finally pardoned in 2006, a good 300 years after her trial.

the tent. They "learned" by shadowing other doctors. According to medical lore, William Smellie, one of the most famous eighteenth-century man-midwives, dressed up like a woman with a frilly cap and dress just so he could see what he was doing. Unlike the sixteenth-century doctor-in-drag who was burned at the stake after his costume was revealed, Smellie lived to tell the tale.

Seventeenth-century delivery room etiquette worked well for the Chamberlens, who were known to sneak their instrument from their medical bag under the tent of sheets so no onlookers could steal a glimpse, not even the woman in labor. According to medical lore, they arrived in the woman's home in a special carriage carrying a gargantuan wooden box with elegant carvings (think coffin, not medical bag). It took two men to lug it into the home. The hoax led patients to believe that they had lots of large contraptions. Next, the Chamberlens kicked out all the relatives. (Most doctors welcomed the companionship of sisters of the parturient woman's mother to calm the laboring mother-to-be.)

> "I will not take apology for not publishing the Secret I mention we have to extract Children without Hooks"

Alone, and with the instrument undercover, the Peters would slip the tool under the sheets with the speed of a magician slipping a scarf up his sleeve, lest the laboring woman peek. And there was more. The Chamberlens were not taking any chances on anyone stealing their ideas. They blindfolded the expectant mother. It's also said that the Chamberlens clanged bells, hammers, and chains so that no one would hear a peep from their instrument, the sound of which could have provided a clue to the design.

Colleagues loathed the Chamberlens because they refused to share their secrets. Yet the Chamberlens were no different from their rivals, just more successful. Medicine was viciously competitive, and doctors coveted crucial clues. There were no patents to protect them and no money to be made from sharing ideas. Hugh Chamberlen, one of the many Chamberlen man-midwives, wrote that he could not divulge the family secret because it would hurt his father and brothers. Or as he put it: "My fathers, brothers, and my self (tho' none else in Europe as I know) have, by God's Blessing, and our Industry, attain'd to, and long practis'd a Way to deliver Women in these case without any Prejudice to them or their infants. . . . I will not take apology for not publishing the Secret I mention we have to extract Children without Hooks. . . . I cannot esteem to publish it without Injury to them. . . . I do but inform that the fore-mention'd three Persons of our Family, and my Self, can Serve them in these Extremeties, with greater Safety than others." In other words, if anyone wanted to ensure they'd get through childbirth alive, they could seek out a Chamberlen, but the Chamberlens were not about to teach their methodology. These were the days before people traveled easily. No laboring woman was about to travel across the country for delivery. Unless you were royalty, the Chamberlens were not traveling to you.

Female midwives hated the Chamberlen family because in 1634, Peter petitioned the king to form a Midwives Corporation that he would run. What a ruckus that caused. Midwives flooded the palace with letters denouncing the Chamberlens, claiming that the man-midwives lacked expertise in childbirth, used violent instruments, and refused to treat the poor. Needless to say, his idea never panned out in his lifetime. Years later, English midwives did organize, without any help whatsoever from a Chamberlen. Poor Peter. The Royal College insisted *he* apply for a midwifery license.

Peter was not going to let a bunch of women get the last word. In 1647, he published a diatribe against midwives. The tome, *A Voice in Rhama: or, The Crie of Women and Children*, claimed—as many doctors would after him—that all midwives were ignorant and have wreaked havoc on childbirth.

Eventually, Peter was kicked out of the Royal College because he missed too many meetings, just like his father and uncle who never showed up at the required Barber and Surgeon meetings. When he wasn't delivering other women's babies, Peter was busy making his own. He married twice and had 14 sons and 4 daughters.

"I am certain that no such thing as bringing a strange baby in a warming pan could be practiced without my seeing it"

Hugh, his one son who dedicated his career to man-midwifery, remained in London during the bubonic plague epidemic that decimated the city in 1665 and killed upward of 7,000 people in one week.* His generous public service earned him slight celebrity status, but nothing compared to the notoriety he received after one of the most famous royal baby scandals. In 1658, Hugh defended King James II when he claimed that the queen had given birth to the next heir to the throne. (Without a son, succession would have passed from Catholic king to Protestant son-in-law, William of Orange.) On June 10, Hugh was summoned to the queen's bedside because she had gone into labor three months early. Hugh said that by the time he arrived, a female midwife had already helped the queen deliver a robust baby boy. It did not take a birthing expert to wonder how such a chubby and healthy baby could have been born so prematurely. Gossip ran rampant through the kingdom. It was said that her baby died and the royal family smuggled a healthy newborn

---

* No one knew that rats transmitted the disease. They blamed dogs (and killed them) and water (doctors advised drinking sherry instead). The dead were carted away. Dead dogs adorned the streets. Those who could afford, fled the city— including many doctors.

into the room to secure succession. The new baby was christened James after his dad. But most people called him the Old Pretender, as in pretending to be royal. (As the rumor mill churned, William of Orange prepared to return to England and claim the throne for himself.) Hugh, the ever-loyal royalist, claimed that without a sliver of doubt the baby was most definitely the queen's biological son. "I am certain that no such thing as bringing a strange baby in a warming pan could be practiced without my seeing it," he proclaimed. He gave no explanation to back up his claims. The king, stung by the gossip, insisted that his court promote Hugh's version of events. He demanded the palace display the bloody and discharge-stained sheets to prove that his wife went through labor. It was to no avail. (Could it be that no one dared to look?) William of Orange took over the thrown. King James, his wife, and baby James escaped to France. Not surprisingly, Hugh endeared himself to the new king and queen, getting the job as midwife to William's wife, Mary. She never had children.

In 1670, Hugh went to Paris to sell the forceps design to Dr. Francis Mauriceau, one of the most renowned man-midwives.* Mauriceau made a wager. If Hugh could prove that his famous forceps could deliver a baby safely and quickly, Mauriceau would buy the design for 10,000 crowns.

Mauriceau chose a 28-year-old woman who had rickets (disfigured bones triggered by severe malnutrition) that twisted her pelvis so much that there was no space for the baby to come out. The pompous Chamberlen entered the birthing room and declared "nothing could be easier." (To make matters worse, the woman's

---

* The male midwifery business got a huge boost in France when Louis IX's mistress gave birth with a male midwife from the get-go. They say that Jules Clement, the accoucheur, delivered the baby blindfolded, as per the king's demand. He did so well that he went on to be the midwife to the Spanish royalty.

water had broken four days earlier, so rampant infection had probably already set in.) Three hours later the woman died. They say that Chamberlen ripped her uterus with his forceps and she bled to death. It's also possible the infection killed her. In truth, not much is known about the young pregnant girl who sacrificed her life for the sake of a dare. We don't even know if she agreed to be part of their wager. Mauriceau tried to save the baby by doing a C-section after the mother died. The baby died too.

As for Chamberlen, he went back to England without a sale and with a bruised ego. But he got his revenge, or at least the last laugh. He bought a copy of Mauriceau's latest book, *Diseases of Women with Child*, translated it into English, and added derogatory remarks about the French doctor-author. He made about 30,000 pounds on the sale, benefiting from the lack of copyright laws.

Then he infuriated the Royal College of Physicians by publishing tips for patients to distinguish good doctors from bad ones. Shortly thereafter, the Royal College accused him of malpractice and fined him 10 pounds for allegedly killing a patient. Hugh had induced vomiting and removed blood from a woman who miscarried during her sixth month of pregnancy. She died shortly after the treatment. In truth, his treatment did not stray from standard remedies of the day. The purging was supposed to remove toxins.

Hugh had other schemes to improve England. He proposed a land bank, a national health system, a national street-cleaning program, and a government-sponsored program to prevent the sale of rotten food. None of his ideas were realized in his lifetime. (Britain's national health service started in 1948, some 250 years later.)

Eventually, Hugh moved to Holland, where, as the story goes, he sold his forceps design to an unscrupulous doctor who jacked up the price and sold intentionally defective instruments. Others say that Hugh sold him half the instrument, which is like selling half of a pair of scissors.

Hugh's son, also named Hugh, allowed the forceps design to go public in England in the early 1700s. The younger Hugh also convinced his friend, the Duke of Buckingham, to build a colossal Hugh Chamberlen monument inside the north choir aisle of Westminster Abbey. It stands today, touting the magnanimity of the Chamberlen family: "In return for a life saved at his birth, for health restored, and at last confirmed, EDMUND DUKE OF BUCKINGHAM, placed this sepulchral monument, to a man the most spotless and friendly." It goes on to say that Hugh "repeatedly saved their only heirs from being snatched away from illustrious families, and their eminent subjects from his dearest country" and that a man "of such fortitude and elevation of mind; of a character so prone to munificence and a nature so ingenuous and liberal, that it had easily been supposed his race originated with a noble and ancient founder, although it were not known that he sprung from a remote branch of the Earls of Tankerville four hundred years old."

After he retired, Peter Chamberlen hid the original Chamberlen forceps in a box under the floor of his country mansion, trying to preserve the secret forever. In 1813—years after Peter died—the mother of the young couple who currently owned the home then noticed a crack in the floorboard. She lifted the floorboard and found a box of antique trinkets. It's likely that even if the mother, daughter, and son-in-law knew that a Dr. Peter Chamberlen once lived in their home, they would not have recognized the name or connected the former owner with anything having to do with childbirth. Indeed, when she lifted the box out of the floor, the mother could have thrown it away. But something possessed her to show it to her friend, a retired surgeon. He knew right away that they were onto something big.

The box contained the original Chamberlen forceps, the very tools that doctors, midwives, and women had been dying to get their eyes on since the late 1500s. What they found was that these highly secretive baby clamps looked like two soup ladles attached with a spring. The nifty part about the Chamberlen design, compared with others, was that you could put the spoons in one at a time, nuzzling them around the baby's head. Then you locked the forceps in place. For the amateur medical historian that this surgeon was, the box of medical tools from years past was as if someone gave a paleantologist an authentic *T. rex* skeleton. The discovery of the Chamberlen model unveiled to the world one of the longest kept family secrets.

Since Chamberlen days—actually beginning with them— forceps have riled up the fiercest debates among obstetricians and midwives. American doctors flocked to Europe to train with the cream-of-the- crop forceps teachers. By the seventeenth and eighteenth centuries, the vast majority of doctors, midwives, and pregnant women worried that the instrument caused more **"dangerous substitutes for their own hands"** harm than good. Those who had forceps, including the prominent Drs. William Smellie and William Hunter, rarely used them. *Ladies Home Companion* told doctors to use herbs, not tools. *Female Physician* (a book written for male physicians about treating women, not a book, as the title suggests to a modern reader, for female physicians) suggested herbs or internal manipulation, but not forceps. One midwife claimed that women were "dazzled" by the new tools because they quickly learned that forceps were "dangerous substitutes for their own hands." Indeed, in the tradition of doctors writing poetry as revenge, John Maubray, the *Female Physician* author, penned an anti-forceps poem:

> Kill *many more INFANTS than they* save *and* ruin
> *many more WOMEN than they* deliver
> *Fairly: . . . when they have perhaps wounded the*
> *Mother, kill'd the INFANT, and with violent* torture *and inexpressible*
> pain
> *Drawn it out Piecemeal, they think no* Reward *sufficient*
> *For such an extraordinary Piece of mangled work.*
> *But, in short, I would advise such to practice* Butchery *rather than*
> *MIDWIFERY;*
> *for in that case they could sell what they* slay.

The turning point came in 1817, when Princess Charlotte Augusta, daughter of Princess Caroline and George, prince of Wales (the future George IV), died after a prolonged labor. This was the first highly publicized case to accuse a man-midwife for *not* using forceps. Afterward, their use increased rapidly. Rather than worrying about the consequences of using forceps, as doctors in the past had done, doctors started worrying about the consequences of not using them.

When used correctly and at the appropriate time, forceps save lives. By the early nineteenth century, many doctors were advocating their use routinely to speed deliveries. Doctors were confident, sometimes overly so. Midwives were worried, sometimes overly so. Women were confused, rightly so.

In 1908, Harvard obstetrician Franklin S. Newell advocated forceps to deliver all the babies of the "overcivilized type" of women. He thought middle- and upper-class women were too frail to push. The typical American woman, he said, "brought up with every care and luxury and little to do except to amuse and take care of herself, not infrequently arrives at maturity a delicate weakly specimen, whose nervous organization seems to overshadow and control her whole physical force." In a science journal, he published detailed accounts of three women whose deliveries required forceps not because the baby was stuck but because the expectant mothers could not cope with labor, or, as he put it, because they showed "exaggerated symptoms." One upper-class woman showed an "utter lack of mental control," another "started to become hysterical during labor, " and the third had recently been released from a sanitarium for nervous exhaustion. Newell added facetiously that doctors had two choices: revamp society so that women would grow fitter, or cater to the new breed of frailty.* In his most prescient joke, Newell said that he envisioned elected cesarean sections, when women could opt out of labor altogether.

The newfound popularity of forceps ushered in a short-lived era of all sorts of weird birthing gadgets. The dynanometer measured the force of the forceps against the baby's head, warning the doctor if he was crushing the baby. The *tractions soutenues* (sustained traction) was a crazy French concoction that allowed three doctors to stand far away from the pregnant woman and control the forceps through a series of pulleys. The advantage over the standard forceps? It's not clear, except that it permitted the doctor to stand farther away from the birth canal—sight unseen—so as not to embarrass the parturient woman. The Italians invented the silliest device of all. It was a do-it-yourself forceps, dubbed

---

* George J. Engelmann, a nineteenth-century physician, wrote a book about birth among primitive people. He wrote: "I will describe the most important time of life for a woman . . . among primitive people labor is short and easy . . . for all people who live in a perfectly natural state." He said "they do not marry out of their own tribe or race and the head of the child is adapted to the pelvis of the mother through which it is to pass." As soon as there is any deviation from these natural conditions, trouble results; the more civilized, the more difficult labor. George Engelmann, *Labor among Primitive Peoples: Sharing the Development of the Obstetric Science of To-Day*, 3rd ed. (St. Louis: Chambers, 1884), 7.

Nineteenth-century Italian do-it-yourself forceps. The fad never took off. G. J. Witkowski, *Anecdotes & Curiosites Historiques sur les Accouchements.* (Paris: G. Steinheil, Editeur, 1892), 82, Courtesy of the Malloch Rare Book Room, New York Academy of Medicine Library.

the autotraction, or *forcipe a staffe briglie*. It was essentially the same as the French contraption with lots of ropes tied to floor beams and wrapped around bedposts, except in this case the laboring woman controlled the ropes. With forceps balanced inside of her and the rope looped around the foot of her bed, she grabbed each end of the rope with her hands and presumably jiggled the ends to pull the baby. Their popularity was fleeting.

forceps fashion

Like everything else in medicine, fashions come and go. Forceps fashion, in vogue during the nineteenth century and early twentieth century, has faded in the past few decades. Today, forceps or vacuums are used in slightly more than 10 percent of all deliveries. For some women, forceps seem old-fashioned and scary, which makes doctors worry about lawsuits should anything go wrong. There are doctors who could easily turn to forceps but opt for the C-section, which incurs the risk of surgery but less risk of litigation. Several doctors have remarked that should anything go wrong, it is unlikely to get sued for *doing* a C-section, but you are likely to get sued for *not* doing one. What's more, it is impossible to prove that any tears or anything that goes wrong during labor is *not* the fault of forceps.

Every ob-gyn resident these days learns the nuts and bolts of forceps and vacuum, but they may not feel as comfortable using them because they're not watching their teachers using them frequently. Like all professions, you can read all you want and get a lesson here and there, but you learn habits by mimicking teachers. If your mentor is not grabbing the forceps, chances are you won't either. Forceps, though at times lifesaving, will never become as widespread as in the past. C-sections are the default option when the baby is stuck.

For many women who live in developing countries, surgery is not an option. They live far from hospitals. In poor countries, more than 40,000 women die every year due to obstructed labors. In industrialized nations, women rarely die when a baby is lodged in the birth canal. International health organizations urge midwives in areas far from hospitals to learn forceps and vacuum extraction techniques.

It's been said that if the Chamberlen boys had passed around the how-to-make-your-own-forceps manual, thousands of babies generations past would have been saved. No one will ever know because no one really knows if these doctors saved lives. No one did scientific studies comparing a new method against standard practice. Results from today's randomized trials and placebo-controlled investigations have taught us time and again that new medical gadgets and drugs do not always live up to their claims. During the Chamberlen reign, doctors learned from experience or hearsay. Women told each other that the Chamberlens were the best birth attendants around. That was good enough.

As historians Richard and Dorothy Wertz wrote, doctors used

the forceps, like ammunition, to drive female midwives out of business. Natural childbirthers today see the introduction of the forceps as the beginning of a slippery slope into excessive use of technology. Yet it's too simple to see male intrusion as 100 percent malicious. Doctors were beginning to unravel the mysteries of the uterus and female physiology. Certainly, some of them went after the female midwives just to bolster their own business, but the motives were not completely masochistic. They may have been kicking out midwives and slandering their work, but many of them truly believed they were doing so to save the lives of expectant mothers and babies. As the late Catherine M. Scholten wrote in *Childbearing in American Society: 1650–1850:* "Though their assumptions about women now appear false, these physicians had reached the innovative position that childbearing women deserved serious medical attention. However condescending some of their words sound today and however imperfect their help may have been, physicians were concerned with the welfare of women. . . . Well-instructed assistants to women in childbirth were one sign of the value placed on women in civilized societies."

The Chamberlens introduced their highly touted secret tool more than 400 years ago. It brought fame and fortune. They certainly exemplified the world of fiercely competitive medicine. There are those today who lambaste this family of man-midwives for coveting a lifesaving tool. But the Chamberlens have not riled historians the way a famous antebellum gynecologist would. That doctor, J. Marion Sims, also became rich and famous. He also perfected a technique that would save lives. He was also egotistical. But his legacy would become far more infamous, and his story would spark more heated debates than the Chamberlens could ever imagine.

3

Slave Women's Contribution
to Gynecology

In the years leading up to the Civil War, about ten female slaves in Alabama did more to advance the fledgling field of gynecology than any other women would ever dare to do in the name of science. Whether they did what they did voluntarily is an entirely different matter. All we know is that these young women, all postpartum, moved into a makeshift hospital in Dr. J. Marion Sims's backyard, where they became guinea pigs in a gruesome and prolonged medical experiment. The upshot for their remarkable endurance was a cure for one of the most horrible side effects of childbirth. Dr. Sims used the slaves to figure out how to repair holes in the vaginal wall.

In the days before medicine to speed labor and C-sections to get babies out quickly, women often pushed for days on end. Not uncommonly, the pushing tore the vaginal walls, sometimes creating a hole between vagina and bladder or bowel. In medical vocabulary, these tears are called vesicovaginal fistulas (*vesico* meaning "sac or bladder," *fistula* meaning "hole") and rectovaginal fistulas (*recto* for "rectum"). New mothers with such fistulas leaked urine and sometimes feces into the vagina, triggering infections. These women were horribly uncomfortable, exuded an awful stench, and for obvious reasons rarely left the house. Johann Dieffenbach, a nineteenth-century surgeon, said he pitied the man whose wife reeked and became a "source of disgust" and "object of bodily revulsion to her husband." As Seale Harris says in his biography of Sims, "Such women never died of their affliction, it seemed; instead they lived on year after year in misery, ostracism, and disgust, wishing for death and sometimes committing suicide to achieve it." Before Sims developed the cure, fistulas were not an uncommon injury among rich and poor alike. Molly Drinker, from Philadelphia society, tore her rectum during birth and was no longer "fit to be around other people." A slave with a fistula was no longer fit to work.

Dr. Sims's cure transformed a gruesome consequence of birth into an easy fix. That's undeniable. What continues to vex historians is how to reconcile a man who achieved phenomenal success for himself and his patients by such awful means. In the words of Barron Lerner, a Columbia University physician and medical historian, "one would be hard pressed to find a more controversial figure in the history of medicine than J. Marion Sims." And yet, Sims's story is more than a window into the history of medical experimentation. His climb from small-town doctor to international celebrity mirrors the tremendous changes in nineteenth-century obstetrics as it evolved from a second-rate trade to a respected profession.

"I felt sure I was on the eve of one of the greatest discoveries of the day."

James Marion Sims was born January 25, 1813, in Hanging Rock Creek, "a straggling country village amid the red hills of Upper South Carolina." He was small and slight with delicately chiseled features. He hated school, nearly jeopardizing his college diploma when he adamantly refused to write five mandatory English assignments. He convinced his professor to reduce his requirement to two papers and then got two friends to write them and forge his signature.

Though his father insisted he become a professional, Sims pleaded for a job in the family's general store. He needed money quickly to prove to his rich girlfriend's mother that he was a breadwinner. Besides, Sims did not think he was smart enough to be either a lawyer or a minister, the two jobs his father had in mind. The only other profession was medicine, which was considered a lowly occupation.

"My son," Sims recalled his father telling him, "I must confess that I am disappointed in you. If I had known this I certainly never would have sent you to college. This is a profession for which I have the uttermost contempt. There is no science in it. There is no honor to be achieved in it, no reputation to be made and to think that *my* son should be going around house to house through the country with a box of pills in one hand and a squirt in the other to ameliorate human suffering is a thought I never supposed I should have to contemplate."

The senior Sims had a point. The local family practitioner was an alcoholic, and the others were not much better. They worked by trial and error, treating the sick with a collection of herbs and pills of questionable merit. If patients recovered, chances were it was in spite of the therapy. The sick were purged, bled, and cupped. Mercury was doled out until patients started salivating, a sign of toxicity. Doctors pushed drugs until patients had wicked side effects. They thought it proved the medicine kicked in, which in a way it had.

Sims's wayward trajectory into medicine was typical for novice physicians. He apprenticed with a town doctor and attended the Medical College of South Carolina for three months. There were no statewide or nationwide requirements. Formal schooling was an option. You could attend for as little or as long as you wanted. Sims attended the esteemed Jefferson Medical College in Philadelphia for a few years. Admittedly, he barely stayed awake during some of the medical lectures, particularly the few talks about women's diseases that made him "shudder inside." When he hung his shingle, he felt helpless. Besides the fact that he rarely paid attention in class, he had had "no clinical advantages, no hospital experience, and had seen nothing at all of sickness."

One lecture kept him awake, not because of his keen interest but because he found it utterly repulsive. The professor told the students how to repair a twisted uterus: Place the woman on her knees and elbows and put one finger in her rectum and another in her vagina and push and pull until the womb rotated back into place. Years later, Sims would heed that advice to help a woman and in the process of curing her devise his world-famous technique to fix vaginal fistulas.

Sims was not the only male medical student queasy about gynecology. It was considered a "messy and unscientific" field. Why would a man with a formal education choose to do the same thing that uneducated illiterate women do? Up until 1886 in England, where medical education was considered the best in the world, medical students were not required to study obstetrics. Delivering babies was a sideline, not a medical specialty. There were the rare male midwives that finagled their way into the royal birthing rooms, but for the most part women preferred women delivering them; and male medical students preferred other specialties that held greater esteem among colleagues. In the early 1800s, a British blacksmith and a grocer started delivering babies to make extra money. Why pay a doctor when you seemed to get the same care for less from the corner grocer?

Sims's medical career did not get off to an auspicious start. His first two patients died. They were babies with severe diarrhea. He later admitted that he had no clue what he was doing with either of them. While the parents waited frantically, he flipped through his medical book trying one concoction of herbs after another. He cut the baby's gums "down to the teeth" to relieve pressure, as one medical book suggested. Nothing helped. Not surprisingly, after his first patient died, he was terrified to treat the next baby with diarrhea. This time he started at the back of the book and worked his way forward. That didn't help either.

With his first two patients dead, Sims had a hunch his chances of luring more customers in town were nil. He took down his shingle and headed west to Mount Meigs, where he worked with a soon-to-retire doctor. His first patient was a woman dying of tuberculosis. The elder physician suggested bleeding. As Sims would remark years later, the elder doctor said, "Now she will be better," or as Sims remarked, the kind of "better that comes with death."

In time, a few patients survived Sims's care and boosted his credibility. Motivated by his newfound success, he and his wife Theresa moved to Montgomery, where he started, as he said, from treating those at the lowest ranks of society and slowly moved upward. That meant treating "the free niggers," then Jews, "very clannish . . . but always had money in plenty, and were liberal with it" and eventually he saw a few "upstanding" families.

Montgomery was his watershed. He claimed to successfully treat a woman with a disfiguring cleft lip and to cure newborn lockjaw by pushing against the baby's skull. (Lockjaw is the common name for tetanus, caused by the germ *Clostridium tetani* and triggering muscle spasms and convulsions. The shoving did nothing.) He also claimed to fix a facial tumor, hoodwinking a slave into a barber chair, where he restrained him with leather straps and then used a chainsaw to hack off a tumor. He "appeared very much alarmed," Sims recalled. Most importantly, he realized he liked to cut. His enthusiasm for the knife would last his career. He would become a proponent of surgically removing the ovaries to cure neuroses and slitting the cervix to boost fertility.

But it was one patient who changed the course of his career from general practitioner to woman's doctor. As Sims described it, stout Mrs. Merril fell off her horse, hollering about stabbing pains in her bladder and rectum. "If there was anything I hated, it was investigating the organs of the female pelvis," said Sims. He told Mrs. Merril to get on her knees and elbows, following the advice of his teacher from that lecture of years ago. Still reluctant to have to do what his teacher had instructed, he put a sheet over his patient to protect her modesty and put a few fingers in her vagina but couldn't bear to put another finger in her rectum. While wiggling his fingers around aimlessly, Mrs. Merril exclaimed, "Why, doctor, I am relieved."

Sims assumed that a gush of air into the vagina forced the womb back into place. His instant success energized him. He had an idea. He suspected that wiggling his fingers the way he did inside of Mrs. Merril would enable him to see vaginal tears and in turn figure out a way to sew them. Just the other day, he had turned away patients—slaves brought to him by local landowners—because there was nothing to be done for them. But after the Mrs. Merril victory, Sims started "ransacking the county," collecting slaves to embark on a "series of experiments to the end." Sims was not one to underestimate his abilities. As he wrote, "I felt sure I was on the eve of one of the greatest discoveries of the day."

Sims had treated many female slaves and was confident he could borrow or buy a group of women for his experiments. Plantation owners were eager to help Sims because they wanted fertile slaves. Ever since the 1808 ban on importing slaves, landowners looked to breeding them to provide help to the next generation. (African

men and women were forced to mate in captivity, like animals.)
Fertile women brought in higher prices on the auction block.

Sims was not the first person to do vaginal-tear experiments on
slaves. But he was the first one to do it successfully and the first
one to turn his procedure into global fame and fortune. Fifteen
years before Sims did his trials, Dr. John Peter Mettauer tried the
same kind of thing on several slaves, but he used lead sutures
instead of silver. He said it worked on two of three women.

Sims was also far from the only Southern doctor to exploit
slaves to further medical research. Dead slaves were coveted by
medical schools. One flyer proclaimed: Our "slave population
of the city, and neighboring plantations, is capable of furnishing
ample materials for clinical instruction."

Medical treatment for slaves was sold under the guise of free
care, but often it was dubious treatment under the direction of
inexperienced medical students. In the best of situations, poor
women sometimes got superb care from top-notch doctors. The
catch was that there was usually an audience. The questionable
part was the term *care* and how often these desperately indigent
and powerless folk were truly *asked* to participate in experiments.
There was no such thing as informed consent—or any kind of
consent, informed or otherwise.

In *Medicine and Slavery*, Todd L. Savitt describes a series of
reprehensible trials and treatments targeting the black population. Among the many, Dr. Walter F. Jones promoted a remedy
for typhoid pneumonia that involved pouring buckets of boiling
water on the patient's back. He claimed it promoted capillary cir-
culation. Another doctor made a slave sit over a fire pit to study
how intense heat fried the skin. Some of the experiments were
not outright dangerous, but because of the potential for toxicity
were tested among blacks before whites. Thomas Jefferson, for
one, tested out the new smallpox vaccine on some 200 slaves dur-

ing the summer of 1801. Fortunately, it worked. "The attitude of physicians," wrote Savitt, "seems to have been that blacks, being property, or free but inferior could be used for medical display and publicity, whereas whites with similar anomalies required privacy and anonymity."

The Southern doctors did not have a monopoly on exploiting the vulnerable and voiceless. While they abused black slaves, Northern doctors used poor Irish immigrants. Anyone could pay for a standing-room-only spot to watch doctors do all sorts of weird things to naked poor women. Medical students could request front row and help holding the legs apart for better viewing.

Sims's slow rise to stardom began with Betsey, a teenage black slave who, in the words of Sims, "willingly consented" to climb on a small table stark naked, just big enough for her to get on all fours. No blanket for Betsey, the way Mrs. Merril was covered. But he did put a sheet over the table, probably to keep the furniture clean. Medical students spread her legs and opened the lips of her vagina while Sims used his newest invention, the speculum, made from two large spoons he picked up at the local hardware store. With Betsey mounted on the table and legs pushed apart, Sims exclaimed that he saw "everything, as no man had ever seen before."

"everything, as no man had ever seen before"

Sims could not contain his glee. "The walls of the vagina could be seen closing in every direction; the neck of the uterus was distinct and well-defined, and even the secretions from the neck could be seen as a tear glistening from the eye, clear even and distinct as plain as could be." His buddies from the old neighborhood may have been journeying into the Wild Wild West, but this Southerner claimed he was venturing into territory few pioneers dared to explore. "Fired with enthusiasm by this wonderful dis-

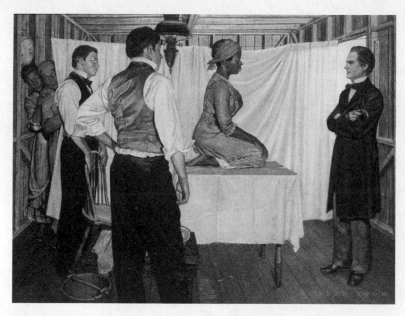

J. Marion Sims with Anarcha, Betsey, and Lucy and two physician partners, illustration by Robert Thom. Robert Thom's image of J. Marion Sims, courtesy of the Collection of the University of Michigan Health Science System, Gift of Pfizer Inc. UMHS.30.

covery," he would later write, "it raised me into a plane of thought that unfitted me for the almost for the duties of the day."

From 1845 to 1849, Sims operated on Betsey, Anarcha, and Lucy, plus about seven other girls, over and over again. He borrowed the girls from other owners or bought them. He sewed Anarcha upward of 30 times. No one knows how many operations the other girls endured, but he did not keep them unless they were preoperative candidates.

He made a lot of mistakes along the way. Once, he shoved a sponge into one woman to absorb the urine but the sponge triggered infection and she nearly died. He admitted afterward that it was a "really stupid thing of me to do." Sims refused white patients because the pain was too much for them to bear. Eventually, Sims achieved his goal and would perfect the repair of this hideous birth complication. He had other medical accomplishments. In addition to the speculum, he created the so-called Sims position: placing a woman in a fetal position to make it easier to look inside.

Historians have attacked Sims because he refused to use available anesthetics. They said it was another sign of racism. There is no denying that he thought that black people were inferior. He believed, as did so many others, that African-Americans had a higher threshold for pain compared with civilized Caucasians. But that is not why he refused ether. Sims was among many experts who didn't touch anesthetics when they were first discovered, regardless of the color of the patient. They worried the drugs were deadly. A white woman, Mrs. H., asked for the cure for vaginal tears, but he refused because it was too painful. He knew she could not bear his surgery, and he refused to give her anesthesia.

In the beginning, Sims was luring crowds of medical students and doctors to watch and to help. After a year of one failure after another, Sims's enthusiasm did not wane. But his audience did. People stopped attending his performances. No one showed up to help keep the girls' legs apart. The girls who could get out of bed became his helpers. Local physicians were disgusted, not because of what he did to the teenagers, but because he seemed to be wasting his career. "It is unjust to continue this way and carry on this series of experiments," a fellow doctor told him. "You have no idea what it costs you to support a half-dozen niggers, now more than three years, and my advice to you is to resign the whole subject and give it up."

Despite his early failures, Sims claimed that he knew from the start that he would eventually figure out how to repair the tears, and when he succeeded, he boasted that he made "perhaps, one of the most important discoveries of the age for the relief of suffering humanity." As for the slaves, no one knows much about them. Most likely, they were illiterate, so their version of events is gone forever. We will never know whether they were thankful for the cure or angry for the years of abuse. Chances are, if they got better they were sent back to their owners in working order.*

Sims's career skyrocketed as well as his lifestyle. He moved to New York City, became president of the American Medical Association, and founded the Woman's Hospital of New York. He fled the United States during the Civil War and moved temporarily to Europe to hide his Southern sympathies, which he worried could hurt his business.† Overseas, he promoted his vaginal cure, became man-

> There "was a rapture in his work like that of a lover's pursuit."

---

* What should be a disease of the past, thanks to Dr. Sims, still strikes millions of women worldwide. According to rough estimates, some 3 million women today suffer with untreated vaginal fistulas. These women are poor. They live in countries without proper prenatal or birthing care. They had long labors, often without any midwife or doctor. There were no hospitals nearby to repair their birth injuries. According to rough estimates, every year another 100,000 women suffer this miserable birth consequence, one that has an easy solution but is not accessible to so many women. (L.L. Wall, "Obstetric Vesicovaginal Fistula as an International Public-Health Problem," *The Lancet* 368, no. 9542 (2006): 1201.) There are a few saviors here and there, making their own small efforts—surgeons providing free care or establishing vaginal repair centers. But as many journalists have pointed out, a huge undertaking is needed—money and a solid health infrastructure—to provide widespread care to every woman who needs it.

† Sims was worried that New Yorkers would hate him because of his treatment of slaves. But as Debby Applegate explained in her Pulitzer Prize–winning book about Henry Ward Beecher, *The Most Famous Man in America* (Houston: Three Leaves, 2007), New York City was "schizophrenic" when it came to slavery. Though New York state was home to the first abolitionists, many New York City financiers depended on the slave trade to stay afloat. Applegate said that more

midwife to Napolean III's wife, the Empress Eugenie, and quickly attracted a host of other royal clientele.

He collected a slew of awards but was a controversial figure in his day because he was pugnacious and power hungry. He argued with any doctor he suspected of stealing his glory. He infuriated the all-female Board of Directors at the Women's Hospital, the one he founded, in part because he invited crowds to view his operations. They insisted that patients should not be entertainment for the masses, but they finally agreed to limit the viewing audience to 15. Ever the egomaniac, Sims insisted that his sons legally change their surname from Sims to Marion-Sims to immortalize his name. None of his sons had sons, so the name Marion Sims died along with them.

Sims battled illnesses his entire life, from malaria-like infections when he worked in the South to pneumonia when he moved up North. There were times when he had to take weeks off work because of incapacitating illness. He died in 1883 at the age of 70 after another bout with pneumonia. His death was followed by a festival of memorial services and glowing obituaries in medical journals and national newspapers.

There "was a rapture in his work like that of a lover's pursuit," said *Harper's Weekly*.

"Among the greater luminaries that adorn the professional firmament, Sims appeared as a comet, leaving a path of light," said the Medical Society of Washington, D.C.

"How blessed and sweet were the gentle hand and kindly heart

---

than $200 million poured into the city from Southern cotton plantations. Slaves were used as collateral. A local editor told the London *Times* that if slavery were abolished, "the ships would rot at the docks, grass would grow in Wall Street and Broadway, and the glory of New York, like that of Babylon and Rome, would be numbered with the things of the past." (p. 220) of the Great Benefactor of Woman," said Mrs. Russell Sage upon the commemoration of a statue in Bryant Park in Manhattan.

"Sims was a man of noble character . . . who closed his life without spot or blemish," said a doctor speaking in 1949 at the Woman's Hospital.

One of his earlier biographers called him "one of a few outstanding nineteenth century pioneers who added more to the basic knowledge of medicine and surgery in three or four decades than had been accumulated in all the thousands of years preceding."

There are still portraits and statues of Sims around the East Coast and the South. A large monument stands today, on 103rd Street and Fifth Avenue, right outside the New York Academy of Medicine. Sims looks handsome with his overcoat open, one hand holding what seems to be a rolled scroll, or parchment. He looks like he is about to walk away.

The inscription says:

> Surgeon and Philanthropist Founder of the Woman's Hospital, State of New York. His brilliant achievement carried the fame of American Surgery throughout the entire world. In recognition of his services in the cause of science & mankind. Awarded highest honors by his countrymen & decorations from the governments of Belgium, France, Italy, Spain & Portugal.

And yet, the famed doctor is no longer seen as an icon, but a poster child for patient abuse. Many historians want his story retold to emphasize the slave atrocities that have, for too long, been buried under his achievements. How do you regard a man who performed such freakish experiments on women, but still made great strides for women's health? Is he a hero or a villain?

Do we hold a nineteenth-century Southern slaveowner to twenty-first-century ethics? How would Anarcha, Betsey, and the others tell his story? What about those statues? Is it time to tear down the monuments or edit their inscriptions?

"Hideous as the accounts of his surgery may appear to sensitive twentieth-century eyes, undoubtedly Sims was at least partly motivated by a desire to improve the lot of his slave patients," wrote Caroline de Costa, an obstetrician and gynecologist in Queensland, Australia. Sims was a "dedicated and conscientious physician who lived and worked in a slaveholding society," wrote Lewis Wall, founder of the Worldwide Fund for Mothers Injured in Childbirth. "The operations carried out by Sims on black slave women from 1845–1849 represented his attempt to cure them of an odious and devastating condition that was then considered incurable." But there is the other side. Harriet Washington, in *Medical Apartheid*, pokes fun at the "ethical balance sheet" that tries to rationalize the "savage medical abuse of captive women on one hand and countless women saved from painful invalidism on the other."

Maybe, as a few scholars have suggested, it's not time to take down his statues and erase his chapter in medical history, but enrich the story by building new sculptures to commemorate the women who helped Sims. Sims loathed sharing credit with other doctors, but he did not mind giving his subjects credit. He told colleagues at a medical society that he appreciated the "indomitable courage of these long-suffering women, more than to any one other single circumstance is the world indebted for the results of these persevering efforts. Had they faltered, then would women have continued to suffer from dreadful injuries produced by protracted parturition, and then should the broad domain of surgery not have known one of the most useful improvements that shall forever hereafter grace its annals."

Part Two

4

Dying to Give Birth

Maternal Mortality into the Twentieth Century

When Bertha Van Hoosen was a new doctor on the maternity wards in Detroit in the early 1800s, she saw so many sick postpartum women that she had a recurring nightmare. She dreamed of four coffins leaning against her hospital cot, each inscribed "Doctored by Bertha Van Hoosen." In real life, she said she "had been on duty two weeks before it dawned on me that every patient in the hospital had childbed fever! I had to face the grim fact—an epidemic of puerperal fever!"

Childbed fever was never the leading killer of women, but it was the most tragic. Nineteenth-century death records were unreliable, if existent at all, but sporadic epidemics in maternity wards or in villages injected a constant low level of anxiety among pregnant women.* Husbands carried home babies and buried wives.

---

* In mid-nineteenth century, only some U.S. states published vital statistics (Massachusetts started in 1842, New York in 1847, and New Jersey in 1848. The statistics were not combined until 1880, when these states were added to 19 cities from other states, which totaled about 17 percent of the population. In 1837, England and Wales started a vital registration, but causes of death were not listed

A generation earlier, birthing experts blamed bad spirits or angry thoughts. In the latter years of the nineteenth century, the new emphasis on the science of the female body prompted women to ask doctors for advice and encouraged doctors to look for answers in the lab. They sought solutions in the newly emerging fields of physiology, endocrinology, and bacteriology.

Charlatans, who enjoyed a short-lived freedom to make whatever claims they fancied, targeted women's fear of childbirth by peddling a trove of over-the-counter health cocktails. Dr. Townsend's Compound Extract of Sarsaparilla fixed the "suffering attendant upon childbirth." Constitution Life Syrup was the "most effectual medicine ever discovered" for relief of anything to do with pregnancy and childbirth. Mrs. J. Haskins, a Chicago housewife, claimed that Lydia E. Pinkham's Vegetable Compound made childbirth so much easier. "While I loved children," she was quoted as saying in a newspaper advertisement, "I dreaded the ordeal. It left me sick and weak for months after, and at the time I thought death was a welcome relief." After drinking Pinkham's tonic for the next pregnancy, she felt "strong in health, hardly an ache or pain." Perhaps the wacky health potions gave women confidence about their journey through pregnancy and childbirth, but they did not seem to make any noticeable improvements in maternal health.

Doctors called childbed fever puerperal fever (*Puer*, Latin for "child," and *parus*, "to bring forth"). It was a fever linked to bringing forth children. For centuries, before the age of bacteriology,

---

until 1874. For detailed information on the history of vital statistics, specifically with reference to maternal mortality, see Irvine Loudon's *Death in Childbirth: An International Study of Maternal Care and Maternal Mortality 1800–1950* (Oxford: Clarendon Press, 1992).

that was all that was known. As Oxford historian Irvine Loudon described it, you could deliver a baby Monday, feel fine Tuesday, feverish Wednesday, delirious Thursday, and die on Friday. Childbed fever killed women delivered by midwives and women delivered by doctors. It killed activist Mary Wollstonecraft Godwin in 1797 when she gave birth to her daughter, Mary Shelley, the

Childbed fever is "more criminal than crime."

future *Frankenstein* author. (It's been said that Godwin insisted on a female midwife, but when the baby was stuck in the birth canal, her husband demanded a male midwife, who was later blamed for infecting and killing her.)

Typically, fever struck within days of delivery, followed by sharp pains that radiated from the belly upward. Autopsies revealed thick fetid pus suffocating the ovaries, uterus, and abdomen. New mothers rotted away. Before bacteria were discovered in the late 1800s, doctors were utterly baffled. How could healthy women die so suddenly? It was as if the birth itself consumed them. Not everyone who spiked a fever after delivery died. But there was no way to know who was going to catch the illness, who would survive, and who would go from delivery room to morgue.

There was no known cause or cure but a bizarre assortment of theories. In 1836, a doctor writing in *The Lancet*, a British medical journal, blamed rotten breast milk that leaked downward rather than out the nipples, because he said dead mothers smelled like rotten milk. Others blamed constipation, anxiety, or wafts of cold air gushing into the open cervix.

Women were given laxatives and emetics to drain toxins; chloride douches to cleanse the birth canal; and leeches to suck out bad blood. (Sometimes the bugs crawled up the vagina and got lost in the womb.) At the Boston Lying-In, doctors gave women excessive doses of quinine to fight the fever until their ears rang "like a never-silent church bell." (Ear ringing is a sign of quinine overdose.)

Some women survived despite the remedies. Survival was taken as a sign that the treatment worked. In 1873, a 27-year-old woman with fever tried every known remedy and then some. The experts gave her laxatives, emetics, quinine, and douches. They starved her—and apparently intoxicated her—by limiting her intake to water, tea, broth, and whiskey. She was not allowed to move, which made her ankles swell. She developed blood clots.* After 41 days, this skinny, dehydrated, drunken woman recovered. Her doctors reported her successful battle against childbed fever in a medical journal.

One of the weirdest treatments was fresh-air therapy, a notion borrowed from tuberculosis sanitariums. At a New York City maternity hospital, sick women were carted up to the roof to air out their genitals. Doctors boasted that a postpartum woman came to the hospital skeletal and feverish, but after a month on the roof, she went home cured.

Sporadic epidemics of childbed fever prompted massive hunts for clues. Doctors examined dead mothers, cutting them open and examining the bodies immediately after death. Medicine was not specialized. Obstetricians did not send the bodies to pathology. They could do autopsies themselves if so inclined. Sometimes doctors went from autopsy to delivery, back and forth—death to birth, birth to death, chauffeuring germs with them. Long before doctors eyed bacteria or knew anything about them, a few scien-

---

* Doctors named swollen, blood-clotted legs of immobile postpartum women "milk legs" because they thought they ballooned from breast milk flowing all the way down to the feet.

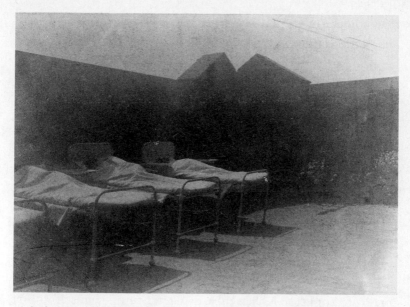

A futile nineteenth-century attempt to cure childbed fever by leaving postpartum women on the hospital roof to "air" out. Courtesy of the Medical Center Archives of New York-Presbyterian/Weill Cornell.

tists suspected contagion. They did not know what was spreading but had a hunch something awful was passing from one sick new mother to the next. In 1773, British obstetrician Charles White made his students wash their hands to prevent puerperal fever. Two years later, Dr. Alexander Gordon, of Aberdeen, Scotland, a brilliant medical sleuth, trailed midwives who had high rates of puerperal fever and compared them with midwives with low rates. He infuriated the midwives and scared the women of Aberdeen when he forecast which pregnant women were likely to be the next victims. His study was perceived more as an indictment than helpful strategy. Similarly, Dr. Oliver Wendell Holmes (father of the Supreme Court justice) called childbed fever "more criminal than crime. The hand that is relied upon for succor in the painful and perilous hour of childbirth becomes the innocent cause of her destruction."

Meanwhile, another doctor was making similar claims against his colleagues. But unlike Holmes, a Harvard professor with clout, Dr. Ignac Semmelweis was a cantankerous newcomer. As Sherwin Nuland tells it in *The Doctors' Plague*, Semmelweis rankled his colleagues in Vienna when he announced that more women died in the maternity wards where doctors worked than in the wards where midwives worked. Semmelweis adopted childbed fever as his life's work. In addition to collecting data—and continually infuriating the medical staff—he performed autopsies on as many dead postpartum women as he could.

Personal tragedy sparked a brainstorm. His friend and mentor, Professor Jakob Kolletschka, died after cutting himself during an autopsy of a woman who died in childbirth. Somehow Semmelweis mustered up the wherewithal to perform the autopsy of his teacher and observed that Kolletschka's corpse resembled the pus-filled bodies of dead maternity patients. Semmelweis concluded that Kolletschka must have infected himself with a deadly substance that killed the pregnant woman. Then Semmelweis reasoned that doctors who perform autopsies before delivering babies carried something deadly from the morgue to the delivery room. He was proposing a sort of germ theory without knowing anything about germs. His colleagues saw otherwise. They saw a young arrogant doctor claiming that doctors were murderers.

Semmelweis demanded that everyone scrub with chlorine after autopsies. He said that washing would keep potentially infected material from passing from corpse to mother. For a while, colleagues adhered to his seemingly over-the-top cleanliness rule and maternal mortality plummeted from 20 percent to 1.3 percent. Skeptics were not convinced. In fact, they were annoyed with the rigid hand-washing rules, so they became lax and, not surprisingly, maternal mortality rose again.

The outrage from colleagues ignited his zeal. He wanted to make rigid hygiene a law. It didn't happen. Semmelweis was fired. He left the wards where he made his discoveries, returned to his hometown, and in 1859 published his manifesto *The Etiology, Concept, and Prophylaxis of Childbed Fever*, in which he called his colleagues murderers. Semmelweis died on August 13, 1865, in an insane asylum. According to medical lore, his enemies—and there were lots of them—drove him crazy. But that is not what happened. Doctors made Semmelweis angry, not mad. Nuland suggests that Semmelweis's madness was caused by Alzheimer's disease. Others point to late-stage syphilis. A few thought he died of the same germs that caused childbed fever, which made for a good well-rounded tale but probably not true. What is known is that Semmelweis was committed involuntarily and died after he was beaten by the caregivers.

It's hard to imagine in today's age of "laudable pus" germ paranoia that there was ever a time that the notion of bacteria and viruses was considered so much mumbo jumbo. The germ theory provided one reason for illness, but it was at odds with several other well-rooted theories. In 1885, Donald McLean, a professor of surgery in Chicago, told students that the germ theory was nonsense. When he saw women oozing with pus after birth, he called it "laudable pus," a normal healing process. Around the same time, a New York City doctor proposed five kinds of childbed fever, only one caused by a germ. He said constipation or hard breasts also cause sickness.

The germ theory was a completely different way of thinking about sickness. It meant that an invisible creature prompted human suffering. Most doctors, in the early years of the germ discoveries, were not looking for single causes the way we do nowadays. They

considered the person's life as a whole to figure out what made
the patient vulnerable. Was she not eating well? Was she worried?
They considered "excitable" or external causes, such as filth or
bad air. German chemist Justus von Liebig promoted the "zyme"
theory that blamed chemicals in rotting vegetables and carcasses.

In *The Great Influenza*, John M. Barry tells the story of Max von
Pettenkofer, who could not believe that one little germ, all by itself,
caused cholera. He believed in the germ theory but thought there
had to be other triggers, too. To prove himself right and Koch (the
discoverer of the cholera germ) the fool, he and several of his students drank cholera-infested water. Shockingly so, they survived,
though a few students got sick. Pettenkofer rejoiced, but not for
long. In 1862, cholera struck two German towns, Hamburg and
Altona. Altona cleaned the water, filtering out the germ. Hamburg
did not. More than 8,000 people died of cholera there. Everyone
blamed Pettenkofer. He committed suicide.

After Semmelweis's death, his life work was vindicated by three
discoveries: Louis Pasteur spotted bacteria in decaying matter, a
crucial clue that germs could spark childbed fever. Joseph Lister
discovered that carbolic acid destroyed germs lurking in wounds; he
named his cleaning technique *antisepsis*. And Robert Koch devised
a four-step program to prove, without a shadow of a doubt, that
germs triggered illness.* Using Koch's methods, doctors proved
that a virulent form of streptococcal bacteria triggered childbed
fever. Later they would blame other germs, too, including staphylococcus and the bacteria that causes gonorrhea. Simply put,
events during the turn of the twentieth century proved that germs

---

* Koch's postulates include these four steps: (1) The bacteria must be present
in every case of the disease. (2) The bacteria must be isolated from the host with
the disease and grown in pure culture. (3) The specific disease must be reproduced
when a pure culture of the bacteria is inoculated into a healthy susceptible host.
(4) The bacteria must be recoverable from the experimentally infected host.

caused the disease; that doctors passed the infection to patients; and that antiseptic technique (scrubbing between patients) prevented infections.

The discovery of the antiseptic technique meant that preventing germs from infecting women should prevent the illness. Antibiotics, the cure, would not be available for decades. And yet, despite these marvelous discoveries before the age of treatment, women continued to succumb well into the first decades of the twentieth century. About 8 mothers died for every 1,000 births. The numbers do not sound alarming, but with so many women giving birth, chances are everyone either knew someone or heard about someone who died in childbirth. Women died for many reasons, including hemorrhages and misused forceps, but childbed fever—a mysterious and potentially fatal ailment that seemed to strike without warning—continued to ignite the most fear and outrage. Death rates should have plummeted, but they didn't. What was going on? *Harper's* magazine published an 11-page article balking at the apathy of the male-dominated medical system and the public health administration: "That a woman should jeopardize her life—or at least her health—when she gives birth to a child has long been accepted as one of the inexorable laws of nature. And so it happens that while science has been waging a winning fight against diseases such as typhoid fever, diphtheria, tuberculosis, and pneumonia, the great problem of maternal mortality has received comparatively little attention. . . . That so many thousands of women should continue to die and to be invalided for life in this country which boasts of its scientific and humanitarian achievements is a disgrace."

There was more finger-pointing than progress. Doctors blamed midwives. They said that midwives were filthy and ignorant. Midwives blamed doctors and their maternity clinics. They said that doctors were too meddlesome, poking inside women with dirty hands and tools. ("Meddlesome midwifery" became the oft-repeated slur in doctor-bashing articles.) They said that maternity hospitals were overcrowded. Public health officials blamed everybody. With typical political hyperbole, Dr. Josephine Baker, the first director of the New York City Bureau of Child Hygiene, proclaimed that the United States "comes perilously near to being the most unsafe country in the world for the pregnant woman, as far as her chances of living through childbirth is concerned."

"There can only be one source of infection from the accoucheur and from without."

America was not the most dangerous place in the *world*, but it was far more dangerous than many other industrialized countries. In a tally of maternal mortality rates including 20 nations, America scored dead last.*

By the turn of the twentieth century, doctors accepted the idea that germs caused childbed fever. But for the next few decades, many of them could not accept the idea that *they* carried the germs. Many doctors tried to find another explanation that would work with the germ theory but get doctors off the hook. They promoted an autoinfection theory, insisting that new mothers got sick by "infecting or by the spread of an inherently bacteriologically fertile reproductive organs." In other words, women put their dirty hands inside their vaginas. Other doctors purported that the pregnant body grew germs. They named this idea the endotoxin theory—a toxin growing from within. In 1902, Dr. George Shoemaker claimed that constipation during pregnancy fostered the spread of noxious germs already inside the pregnant patient. Others suspected

---

* In 1927, for instance, for every 10,000 births, 65 women in the United States died, compared with 48 in England and 29 in Denmark, according to Irvine Loudon, *Death in Childbirth: An International Study of Maternal Mortality 1800–1950* (Oxford: Clarendon Press, 1992).

that the pregnant uterus changed the physiology of the body and spawned illness. *Harper's* said that black women have high rates of puerperal fever because of "bad habits of hygiene and lack of obstetrical attention."

There was a small yet vocal group of doctors, like Dr. Holmes, who rankled colleagues by publicizing the faults of their profession. "It has been found beyond a shadow of a doubt that neither the *Staphylococcus albus* or *aureus*, nor the *Streptococcus pyogenes* nor the bacterium *coli* commune are to be found in the healthy vaginal secretions of the pregnant woman," Dr. S. Marx wrote to the *American Journal of Obstetrics*. He meant that the woman herself cannot be blamed for the infection and that there is nothing special about a pregnant woman that makes her spawn germs. He reported findings from a study that included 48 examinations of 15 pregnant women. None of the women had streptococcus growing in the birth canal during pregnancy. The one woman who died of childbed fever tested positive for streptococcus after delivery. "There can only be one source of infection," Dr. Marx concluded, "from the accoucheur and from without." Eventually, other studies would make it crystal clear that pregnant women may be more susceptible to colds and flu, but the pregnant womb does not grow its own garden of bacteria. The endotoxin theory died. The only way a woman got childbed fever was by having a contaminated hand or instrument inserted during delivery.

The germ theory was a good thing for women. For centuries, anything that went wrong was blamed on some kind of inherent female weakness: Infectious vaginal discharge could be linked to promiscuity or sex fears. Childbed fever could be linked to anxiety. Germs meant that there was a specific, external cause that had nothing to do with the way the woman thought or her morals. Unfortunately, all the talk of warding off germs with hygiene was simply talk. Women would not stand a better chance of surviving childbirth until doctors discovered germ-killing medicines. We would have to wait for the medicine to kill the illness rather than prevent it in the first place. (As always, our medical system is better at cures than prevention.) In the meantime, doctors were doing what they considered the best medicine. They believed they were saving lives by luring women away from midwives and into the hospital, where doctors could control the business of babies. Ironically, what they thought was the best medical care was sometimes the deadliest.

5

Leaving Home

New York's Lying-In and the Growth of Maternity Wards

With great fanfare and local newspaper publicity, the doors of the new building of the Lying-In Hospital of the City of New York opened on January 22, 1902. The colossal brick and limestone medical center had 1,159 windows, 673 doors, and stretched a city block between 17th and 18th Streets along Second Avenue. Like most maternity hospitals, it was called a "lying-in," an apt name because women were confined to their beds for weeks at a time. The doctors trumpeted that the grander complex would enable them to take care of large numbers of pregnant women and inevitably lower the city's abysmal maternal death rate. At the very least, they hoped the buzz about the reopening of the Lying-In would dissuade women from going to midwives. "There is no single factor which will do more to overcome the deplorable midwife practice than the substitute provided by our maternity hospitals," the medical directors said.

In 1900, only the desperately poor ventured into maternity wards, practice grounds for new doctors. Dr. Bertha Van Hoosen said that her Detroit hospital "received only delinquent girls, but none of them was of the prostitute type." Sometimes postpartum women cleaned the rooms in return for services rendered. Often the patients were not direly poor but were in dire straits. In 1905, a New Jersey housewife wrote to New York's Lying-In, asking them to admit her "lovely, Christian housekeeper" impregnated by the son of one of Princeton's finest families. She offered to adopt the baby "born of good stock." (You can't help but wonder if the maid had a fling with the housewife's son—or husband.)

Maternity hospitals were scary places. During the seventeenth century at the Hotel Dieu in Paris, a world-renowned hospital, three to five pregnant women shared one bed. If a woman died in childbirth, the other ones waited hours or even overnight for the orderly to cart the corpse away. A maternity ward at capacity meant that each bed had its maximum of five women. Then women were sent to other wards, which meant that she could be sharing a bed with a sickly and contagious patient. Not surprisingly, one in five pregnant women died during delivery or shortly thereafter, probably spreading their germs back and forth. It could not have helped the situation that the hospital became a tourist attraction, with passersby, peddlers, and beggars wandering the wards, ogling at the women.

Hospitalized pregnant women had no idea that all too often their care was left to the whims of medical students with supervising doctors far away. Most medical schools required students to shadow a senior student a few times before delivering a baby on their own, but the mandate was rarely enforced. Many students never found the time, and then all of a sudden they were delivering a baby without ever having watched anyone else do it. According to medical school lore, a new mother once said to her doctor, "I hope I did not behave too badly. You see, this is the first baby I've ever had"—to which the doctor replied, "You've noth-

ing on me. This is the first baby I've ever seen born." According
to another story that sounds more parable than fact, a Harvard
student dropped a newborn baby twice and then said to the baffled parents, "Sometimes we have to drop them as many as three
times to start them breathing properly." When Van Hoosen was a
medical student, she was completely unprepared to stitch a woman's torn vagina after birth. She grabbed her
medical textbook and read as she sewed, **"Strange as it**
having never seen the procedure before. "As **may appear . . .**
luck would have it," she said, "the surgery **fewer women**
went well." **die when**

The birth of New York's Lying-In Hos- **treated in their**
pital, one of America's largest and most **own squalid**
respected maternity centers, is emblematic **ill-ventilated**
of the way all hospital births evolved in
America. Though disdained by middle-class **houses and**
women initially, the popularity of hospital **nourished by**
births skyrocketed in the first few decades **the coarsest**
of the twentieth century. In 1900, 5 percent **of food than**
of women gave birth in hospitals. By the **when inmates**
1930s, about half of all women and 75 per-
cent of women in cities delivered in hospi- **of the best-**
tals. And by the 1960s, nearly every pregnant **equipped and**
woman chose a hospital birth over a home **best-man-**
birth. The dramatic shift has been portrayed **aged lying-in**
as a successful campaign by obstetricians, a **hospital in**
fairly new specialty at the turn of the cen- **the world."**
tury, clamoring to bolster their status and
push midwives aside. And yet there is so much more to the story.
The increasing lure of the hospital reveals the changing values
among American women, who believed that maternity wards

offered women a safer alternative plus a chance to get out of the
house and be catered to by a crew of trained childbirth experts.
By mid-century, pregnant women would look at home births as a
quaint relic of the past.

The story of the first Lying-In Hospital harks back to the Revolutionary days, when about 4 in 100 women died in childbirth
and when one doctor founded a tiny clinic for women in labor.
The story culminates in the twentieth century after a stormy history that included big names, big dollars, big battles, scandals, and
murder. But more broadly, the story of the Lying-In is a story of
women struggling to survive the birth of their babies and how
doctors and midwives struggled to keep them alive.

The idea for a maternity clinic began in 1798 over drinks at the
City Tavern, a popular haunt for Revolutionary leaders and busi-
nessmen. David Hosack, a well-connected doctor, asked his rich
friends to help him buy a tenement so he could offer health care
to poor pregnant women. Hosack liked to say that he was inspired
by the yellow fever epidemic that left many pregnant women wid-
ows. He assumed that men were more susceptible, which was not
physiologically accurate but was true in a sense. Though doctors
did not know it yet, the yellow fever germ is transmitted by mos-
quitoes. Pregnant women, who tended to stay inside, were probably less likely than their husbands to get bitten by infected bugs.
Still, the real impetus was probably that his own wife, Catherine,
died after childbirth in 1796.

The early years of the Lying-In were a constant struggle for
money. Hosack was determined to keep his clinic independent,
but in 1801, when he could not afford the upkeep, he reluctantly
rented space at New York Hospital (later to become part of the
Cornell Medical Center complex). The union lasted until 1827,
when New York Hospital refused to rent the space for maternity
care any longer. Administrators said they were angry that Hosack

did not share donations. Others suspected that the hospital was afraid of childbed fever epidemics spreading to other wards.

They had a point. The very same hospitals, such as The Lying-In, that were built to prevent and cure such feared consequences of pregnancy often did just the opposite. They transformed what had been a sporadic tragedy of puerperal fever into deadly epidemics. "Strange as it may appear," a local New York newspaper reported, "the results of trials and investigations show that fewer women die when treated in their own squalid ill-ventilated houses and nourished by the coarsest of food than when inmates of the best-equipped and best-managed lying-in hospital in the world."* Doctors had a clever name for hospital-induced diseases: hospitalism. It's a lovely moniker because it removes all blame from the doctors and treats contagious diseases spread through the wards as if they just happened to happen.

Hosack's venture died, but the Society of the Lying-In that he founded continued. It existed in name only. Members collected money and had meetings every now and then. There was no hospital, and there was not much charity work—until, that is, a half century later, when a fortuitous meeting occurred between a doctor seeking funds for a maternity project and the president of this lame society seeking a project to sponsor. The meeting would link the legacies of turn-of-the-nineteenth-century David Hosack with turn-of-the-twentieth-century Drs. Samuel W. Lambert and James W. Markoe. Their efforts to foster the Lying-In would turn

---

* Though most historians have shown that the vast majority of lying-in hospitals increased the risk of childbed fever, Lisa Forman Cody, in an article in the 2004 *Bulletin of the History of Medicine*, claims that at least one London hospital did offer better care. L. F. Cody, "Living and Dying in Georgian London's Lying-In Hospitals," *Bulletin of the History of Medicine* 78, no. 2 (Summer 2004): 309–348. Regardless, the feeling among most women was that if you wanted to survive childbirth, you were better off at home and far away from maternity wards.

the listless society into one of the country's largest maternity centers.

Prior to the meeting, Lambert and Markoe had been running their own maternity clinic. Like many young doctors, they were inspired by a study-abroad stint in Germany, where they had the opportunity to shadow doctors and learn skills not yet taught in America. They learned how to use forceps and how to do craniotomies, cracking babies' skulls during difficult deliveries. When they returned to the United States, they were determined to open their own clinic.

By this time, there were already several lying-ins in America, including the Sloane Maternity in Manhattan, the Boston Lying-In (where Lambert had visited for ideas), and others in Chicago, Philadelphia, and Detroit. Metropolitan Hospital, in New York City, treated women, but it was on Wards Island, a spit of land off the northern tip of Manhattan. In the days of horse and buggies, that was a 3-hour bumpy ride from home to hospital. If you had a choice to stay at home with a midwife or bump around on the back of a horse-drawn carriage in the throes of labor to a clinic run by strangers, you would have opted for the security of your bedroom.

Lambert and Markoe thought they'd do better than the competition in New York by opening a clinic on the Lower East Side, where poor immigrants lived. In 1890, they bought a tenement at 312 Broome Street and used all sorts of clever tactics to fund it. They raised money by producing a show. They also brought in much-needed cash by charging students to take their qualifying exam. Students, then, did not have national medical exams. They could take whatever test they wanted or not bother at all. Shocking to the ambitious doctors, few folks wanted to deliver in their clinic and few medical students wanted their exams or expertise. Most women had midwives, trusted and sometimes experienced women from their home country who spoke their language and provided them with the kind of food they were used to eating.

A lucky break came when Dr. Alexander Lambert, Samuel's brother and medical student, went to check on a woman hemorrhaging after delivery. Despite an array of dubious treatments, she survived. Her family was thrilled. The father wrote a flattering article in *The American Hebrew*, the local Jewish newspaper, raving about the wonderful free care and food donated by Gentiles. "The Jewish clientele boomed by leaps and bounds," cheered Lambert and Markoe. What's more, they had a feeling that the free service, rather than the 8-dollar-a-delivery midwifery service, was "tempting to the thrifty Lower East Side citizen." (Many midwives charged as little as a dollar for a delivery.)

But the real coup was the meeting that Lambert's father had with the president of the Society of the Lying-In. In 1882, Dr. Edward W. Lambert, president of the Equitable Life Insurance Company, was dining at the Lawyer's Club at the same time as William A. Duer, a lawyer and president of the Society of the Lying-In. Lambert asked Duer to help fund his son's maternity clinic, and he agreed immediately, offering $2,000. One thing led to another, and within months, Lambert and Markoe's clinic was absorbed into the Society of the Lying-In. It was a win-win situation: it provided the young doctors financial backing and clout; and it offered the society something to do with their accruing funds.

With the support of the society, business skyrocketed. In the first year of business—in the pre–Lying-In Society years—doctors delivered about 199 babies. By year three, boosted with Lying-In money, doctors and students delivered more than 2,500 babies. What's more, about 360 fee-paying students rotated through the service. Overall, some 591 women suffered from childbed fever but doctors blamed the infections on midwives or the local family doctors, who had cared for the women prior to hospitalization, or

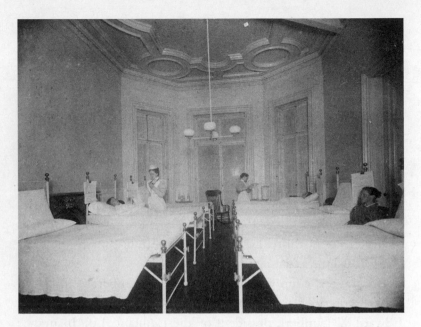

The maternity ward of New York's Lying-In Hospital circa 1897, when the hospital was in the Hamilton Fish Mansion, which served as the hospital building from 1894 to 1899. Courtesy of the Medical Center Archives of New York-Presbyterian/Weill Cornell, a Rockwood photograph.

on the constitution of their patients, who were, according to the doctors, "weak and anemic" Russian and Polish Jews who lived in "squalid, filthy, and wretched" homes.

In 1893, the young hospital got even richer, thanks to financier J. P. Morgan, one of Markoe's patients and soon-to-be close friend. (Markoe became Morgan's personal physician and would join him as doctor in residence during the Morgan family holidays.) As Markoe tells it, one day he brought Morgan to tears when he told him that he had to do a cesarean section in the patient's kitchen because there was no room at the clinic. Morgan responded by reaching into his pocket and handing him wads of cash, announcing, "Get her everything she needs."

Shortly thereafter, Morgan bought Markoe the Hamilton Fish mansion on Second Avenue so the doctors could expand their maternity care. The local press jeered at the Lying-In's new home, saying that the home that used to welcome debutantes would now welcome another "sort of woman." The mansion had enough room for 12 patients and 20 medical students. Mrs. Cornelius Vanderbilt donated the linens. The doctors felt that by convincing women to deliver in their hospital, they would save lives. They were shocked that immigrant women preferred their own squalid homes to the stately mansion-turned-hospital. It was a conundrum facing obstetricians in every major U.S. city. Despite the grand surroundings and fine linens, hospital births tore women from family and friends, from their mother and sisters who often catered to them before and during labor, delivery, and the weeks thereafter. Hospital deliveries meant eating strange new foods. It was not only a move from home to hospital, it was a move from one culture to another, from the familiar to the strange. It meant you were on doctor's turf. At home, as historian Judith

"What is needed is a department in the hospital in question where these cruel parents shall be instructed as to the enormity of the crime they are committing in bearing children in their present condition of ignorance and poverty."

Leavitt tells it in *Brought to Bed*, "if physicians insisted on procedures alien to the birthing women or their families, women sometimes asked them to leave." Not so in the hospitals, where women were on their own, without families to help them weed through the medical morass.

Doctors tried to convince women that doctors were better than midwives by slandering the midwives. Markoe and Lambert, for one, worked the press—and successfully so—by inundating them with midwife-bashing releases. An 1898 article in the *Evening Post*—a Lying-In report word for word—said that nearly half of the "unfortunate mothers were dependent on the care of ignorant midwives. The Department of Health deplores the fact that so large a number of the destitute poor are still dependent on unskilled midwives and believes this unfortunate condition will continue until the present maternity service in this city is either greatly enlarged or some system is devised to care for the neglected cases."

Lambert and Markoe advertised for students in national newspapers. They placed an ad in the *Denver Times* inviting medical students to come to New York for the opportunity to pay $10 for a two-week experience to deliver babies and meet "real rough characters." Students worked 24/7. They were not allowed to leave the building except to make house calls. House staff, the name used today for the most junior doctors, really meant house staff. "As if to inculcate the idea that obstetricians should not sleep much or too comfortably," one student recalled, "we slept in a bed that had been donated, and much too short." He went on to say that the "fuel supply was gathered somewhat after the plan of the Israelites' manna in the wilderness."

The doctors also tried to push through a bill that would restrict midwife practices severely, basically outlawing them. It didn't happen. They used the police to spread the word, inviting a few policemen to visit the clinic one evening. They needed to persuade suspicious authorities that they weren't running an abortion clinic.

The New York Lying-In became a magnet for rich women looking for do-gooder activities—thanks probably to Morgan's interest. Whereas the Brooklyn Maternity Hospital ran bake sales, collecting $80 here, $52 there, the Ladies Auxilliary of the Lying-In hosted gala events at the city's most luxurious hotels, raking in thousands of dollars with a guest list that included Roosevelts, Cushings, Vanderbilts, and Stuyvesants. A few of them visited the postpartum poor. It was an odd schizophrenic response to the immigrant issue, as the ladies raised money to help the poor pregnant women and yet feared their proliferation. As the editors of a 1905 *Vogue* piece said, "What is needed is a department in the hospital in question [the Lying-In] where these cruel parents shall be instructed as to the enormity of the crime they are committing in bearing children in their present condition of ignorance and poverty."

The Ladies Auxiliary continued to bring in money to the Lying-In Society, but true financial security came when J. P. Morgan announced a seismic pledge. He donated $1.3 million to knock down the mansion and erect an eight-story, block-long modern medical complex. Morgan chaired the board of governors, and James Markoe became the medical director. And that's when the troubles began to brew.

"Don't let the public be duped."

On May 20th, 1899, the mansion was demolished. The new building opened for business three years later and was deemed by *Harper's Weekly* a "monument to the generosity of one man . . . to lighten the suffering of 186 women every fortnight during the most critical period of a woman's life." Students traveled from all over America to work with the New York City doctors and live in the hospital's basement for two weeks at a time. Patients came from the city and nearby towns.

Not everyone was happy. People complained that maternity hospitals were built to boost careers, not to help the indigent. A cartoon in the *New York Herald* on May 25, 1897, showed a nattily clad doctor with a large sign around his neck. It said: "Please Help the Poor Young Physician." A letter to the editor of New York's *Mail and Express* asked why the hospital should get government funding. Our tax dollars, he wrote, are not "so much for the benefit of the poor, who do need it, but to a few young doctors." Another letter in the *Evening Post* told people not to be fooled by these money-grubbing doctors. "Charity may incidentally be extended to the starving sick, yet the main purpose of the hospital is not such charity but to afford clinical material necessary for carrying on the business of medical teaching." The letter writers who opposed the hospital said the poor had enough options, with the 45-bed Sloane Maternity, the 30-bed Marion Street Dispensary, and the 150-bed Maternity Hospital, to name a few. As a letter to the editor in the *Evening Post* put it, Morgan can spend his money however he chooses, but "don't let the public be duped." The skeptics may have been correct about the motivations to build the Lying-In, but they were wrong about the hefty supply of medical care for the needy. There were enough poor pregnant women in the city to fill several hospitals if women chose to go that route. In 1900, a baby was born in New York City every 10 minutes.

At the Lying-In, things were looking better on the outside than the inside, where fears were not so much about childbed fever but what the press called "the Markoe Feud." Many of the doctors resented Markoe's arrogance and intolerance. They questioned his abilities and claimed that his quick rise to power was based solely on his friendship with Morgan. On May 11, 1905, nearly every doctor at the Lying-In resigned. "This hospital," the *Chicago Tribune* reported, "the largest of its kind in the country if not the world finds itself confronted with the resignation of almost the entire staff." Only four junior doctors remained. "It was almost impossible for any self-respecting doctor to work in the place," one former staffer said.

> "The mother of hospitals . . . now becomes through association a Hospital for Mothers."

Dr. Markoe's response, as told to the *New York Times*: "Everyone is replaceable."

The hospital became fodder for the tabloid gossip. There were stories about baby switching (that prompted the Lying-In to start tagging babies); about strangers sneaking into the wards to steal newborns and sell them; and about unhappy patients trying to escape. None of the stories were confirmed by reliable sources, but even so, they expressed a lurking fear about childbirth in maternity hospitals. What was really going on inside those closed doors? Why were so many women, too many women, going from delivery table to morgue? Even Dr. Markoe conceded that women were dying at "criminally high" rates. In 1918, after 28 years as hospital chief, Markoe resigned. Two years later, Markoe was murdered while praying at St. George's Episcopal Church. The gunman, an escapee from an insane asylum, barely missed shooting other churchgoers, including such luminaries as Herbert L. Satterlee, J. P. Morgan's brother-in-law, and George W. Wickersham, former U.S. attorney general. Markoe was carried to the nearest hospital, which was, ironically, the Lying-In. He died in the hospital he founded.

In the days following his highly publicized murder, there was much talk that the murderer was an antiwar anarchist because of the high-powered congregation. But the rumors were nothing more than rumors. Or as the church rector told the *New York Times*, "I thought he was some Bolshevik, I was relieved to learn he was an insane and irresponsible person."

In 1928, the fiercely independent Lying-In became part of Cornell University. It was a reunion of sorts, as the two had been connected a century ago. The *New York Times* wrote, "The mother of hospitals, as the New York Hospital has been called, now becomes through association a Hospital for Mothers: and the Lying-In brought with it . . . what might be called a dowry of $6,000,000."

The hospital, like all maternity wards, would become more popular as the century progressed, luring rich and poor alike. What had been places of last resort would change into the top-choice maternity centers, where doctors would offer the latest in hygienic and scientific care.

In the next few decades, doctors would routinely shave women's pubic hair (a closer shave for indigent women, who were considered dirtier) and give them enemas and vaginal douches of bichloride of mercury. To appeal to the wealthier clientele, some maternity centers offered private rooms with lush surroundings. Doctors believed that women needed to be in the hospital to get the best care in the cleanest surroundings. They also knew that a hospital birth spared them from running from house to house. As the renowned Dr. Joseph B. DeLee, an obstetrician at Chicago's lying-in, wrote in the 1920 edition of his medical textbook:

> A distinction between hospital and home practice of obstetrics is gradually creeping into our scientific discussions. Careful study of existing conditions will convince any one that the safest place for the parturient woman is the special, well-equipped lying-in hospital. Here are all the facilities for the aseptic conduct of labor and the puerperium, here is the danger of child-bed infection properly evaluated, here only are the refinements of an operative technic possible, because the operator has the help of trained assistant. . . . another benefit not so generally recognized is the effect on the physician. The maternity relieves him of a great deal of actual labor, it saves him many hours of tedious waiting, it lightens the burden of responsibility, security which reflects itself in his work. The drudgery inherent in obstetric practice is thus largely eliminated and the field becomes inviting to the best men of the profession.

And yet the debates between home versus hospital continued to simmer. In 1938, *Life* magazine ran a center spread called "Birth of a Baby" that showed a woman giving birth at home. (The essay was printed in an easily removable centerfold for parents who deemed the photos inappropriate for their children.)* The photographs, stills from a documentary film, began with a prenatal visit of a woman chatting with her obstetrician and ended with the same woman, a few days postpartum, hugging her husband. The film was produced by the American Committee on Maternal Welfare, made for "the high purpose . . . to reduce illness and mortality among mothers and their babies." The birth photos show a close-up of a baby emerging from a shroud of white sheets, as if it is being pulled out of a tent. According to the captions, Mary, the expectant mother, is giving birth at home. In reality, according to an addendum in *Life*, the woman in the opening and closing shots was actress Eleanor King, who was not pregnant during the film. The newborn scenes that were promoting home births were shot in the delivery room of Cornell Medical Center by a real pregnant woman who asked to remain anonymous.

---

* When I read about the "Birth of a Baby" spread and ordered my old *Life* magazine, my first copy had the note from the editor but was missing the entire essay. Apparently, that subscriber did not want the children seeing it.

6

Birth Is but a Sleep and Forgetting

In 1914, Charlotte Carmody sailed to a mountain clinic in Germany to give birth to her fourth child. Her three toddlers stayed home in Brooklyn, while she spent her last few weeks of pregnancy in spa-like conditions in the mountains of Freiberg. She went because she read a magazine article about a new and easier way to deliver. *Dammerschlaf,* or twilight sleep as it would become known in America, provided women with drugs during labor that lulled them into a woozy forgetfulness during the birth of the baby. As Carmody recalled, "I was given the first treatment at 9 o'clock at night. In small doses the treatment was repeated and I passed into the *'dammerschlaf.'* At 9 o'clock the next morning I awoke, but felt nothing more than a peculiar sensation. I heard my child's voice for the first time and it was brought to me. I had not even known that it was born."

That, according to Carmody, was the best part of the whole thing. She did not remember a whit about giving birth. She went into the hospital one night and the next day a nurse handed her a washed and bundled baby. The birth experience, or lack thereof, transformed Carmody into a crusader for what she and other advocates would falsely claim were pain-free deliveries.

Twilight sleep, popularized by Drs. Bernard Kronig and Karl Gauss, combined morphine and scopolamine, a potentially toxic cocktail that induced analgesia and amnesia, not anesthesia. The distinction is crucial. Anesthesia provides complete pain relief and unconsciousness. Analgesia offers partial pain relief. The added amnesia meant that women felt pain but forgot about it—which makes you wonder if anything is painful if there is no memory. Or as another twilight mother put it, "It makes no difference to me what the doctors say about 'forgetfulness.' For me it was painless. . . . The night of my confinement will always be a night dropped out of my life."

The first two decades of the twentieth century ushered in the seemingly paradoxical age of feminism and flappers. For the first time, as historian Stephanie Coontz tells it, women were going out with men *without* chaperones, bobbing their hair, smoking cigarettes, wearing makeup, and learning the tango. They were buzzed on cocaine-laced cocktails, starving themselves to fit into Coco Chanel's waist-free couture, talking about sex, and having sex. Somewhere between one-third and one-half of college-age women admitted to intercourse before marriage.

Lithuanian immigrant Lane Bryant, nee Lina Himmelstein, started the first line of maternity wear, transforming a dollar-a-week seamstress business into a multimillion-dollar national enterprise. Her stretch-waist skirts offered a comfortable alternative to the corsets that some women wore well into pregnancy—the belly-pinching gear allowed pregnant women to get out of the home and about without "showing." Doctors cheered the baggy clothing because they blamed corsets for crushing reproduc-

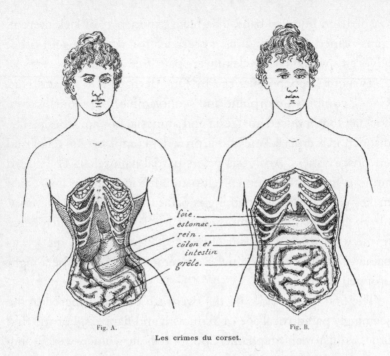

Fig. A.   Fig. B.

Les crimes du corset.

Illustration used to denounce the crimes of the corset. By the early 1900s, doctors warned that corsets, sometimes worn in the first months of pregnancy, damaged internal organs. Courtesy of the Wellcome Library, London.

tive organs and suffocating fetuses. When asked when pregnant women should stop wearing corsets, Dr. Alice Stockman replied, "200 years prior to conception."*

Bryant's clothes suited the fashionably big turn-of-the-century woman. Large, athletic bodies were not only symbols of beauty

---

* Bryant and her second husband, Albert Malsin, tried to advertise in newspapers, but for years no one would print the words "maternity clothes." Finally, in 1911, the *New York Times* ran one tiny notice and her inventory completely sold out the next day. To get around the advertising ban, Bryant launched a hugely successful mail-catalog business. A customer relations guru, she promised to reoutfit any customer whose clothes were destroyed by disaster. And she did. She sent clothes

but considered healthier. Doctors said that well-fed girls did not get morning sickness. The *Ladies New Medical Guide* included a color illustration of the "model of health and beauty," a young woman in underpants with a slight paunch and hefty thighs. Mothers encouraged their college-bound daughters to gain the freshman 15. Another health book cheering the full figured exclaimed: "What more charming sight than a rosy, robust young woman! To such there are no fears, no forebodings in maternity!"

At the same time, women were fighting for the right to vote, the right for birth control, and the right to give birth *their* way. The seeds of reform were planted at the 1848 Seneca Falls Convention, the feminist conference in upstate New York, but the women's rights movement really blossomed in the early decades of the twentieth century. The National Birth Control League and its competitor, The Voluntary Parenthood League, were founded. More than half a million people showed up in Washington, D.C., on the day before Woodrow Wilson's 1913 inauguration to watch "8,000 marchers, 26 floats, 10 bands, and 6 chariots," parading for the woman's right to vote, according to Gail Collins. Mary Ware Dennett was convicted of sending obscene material through the U.S. postal service when she mailed "The Sex Side of Life: An Explanation for Young People." Feminists also pushed for the Sheppard-Towner Maternity and Infancy Protection Act, a federal program that gave grants to states for prenatal and child health clinics. The act also provided funding for nutrition and hygiene education, midwifery training, and home visits by nurses to check on pregnant women. (The American Medical Association, a fierce

---

to 58 customers whose homes were destroyed in a major explosion in Texas in 1947. But the best thing she ever did was expand from clothes for the big bellied to clothes for the large framed, which was probably one of the greatest inventions for pregnant women who can never get off the pregnancy weight.

opponent, claimed that it was a step toward socialized medicine. The act was repealed by the end of the decade. Society ladies joined the progressive movement to host charity events, particularly for causes related to mothers and babies. One star-studded gala to raise money for a maternity hospital included a midnight dinner at the Waldorf Astoria followed by dancing until dawn.

The new drugs for delivery arrived in America at an opportune moment: the first wave of feminism coincided with the rise of the maternity hospital. Women of all classes were beginning to believe that the maternity hospitals offered the safest way to deliver babies. And yet, a few forward-thinking women recognized that hospital births also meant that doctors could call the shots, and they were not willing to go that far. "The women of America are demanding that the administration of painlessness shall not be left to the decision of the doctor, but of the mother," wrote advocates Marguerite Tracy and Mary Boyd in their 1915 book about twilight sleep. Ironically, they considered being knocked out with drugs an expression of feminism.

Tracy, Boyd, Carmody, and others of their ilk—all upper-class feminists—became proselytizers for drugs on delivery. They founded The National Twilight Sleep Association. They went on nationwide speaking tours pushing drugs for delivery, held fundraisers for twilight-sleep maternity hospitals, produced a twilight-sleep movie, paraded their twilight-sleep babies in department stores, and photographed them for women's magazines to prove to skeptics that the drugs did not harm newborns. "We have to accept painless childbirth for the good and comfort of motherhood," Carmody proclaimed to a group of women in Washington, D.C. "We must stand together and insist that it be adopted in this country by the medical profession, and not allow it to be kept from us."

For a while, twilight sleep became the modern way of birth,

particularly among rich women and particularly among rich women who were not afraid to stand up to their doctors. The association's influential membership included birth control activist Mary Ware Dennett, feminist writer Mary Boyd, and Madeleine Talmage Force Astor, widow of multimillionaire John Jacob Astor IV. (They were on the *Titanic* when she was five months pregnant. She survived. He didn't.) For a time, the president of the association was Mrs. C. Temple Emmet, an Astor granddaughter.

"Not a few women of good normal minds have gone to seed, become dumb, patient, brooding animals after the exhaustion of a succession of painful births."

The Twilight Sleep Association illustrates, perhaps more than any other turn-of-the-century organization, the zeal of the early feminists; the simmering anger toward male doctors; and the power of the media to shape public opinion. It was not simply another way to deliver a baby, it was a feminist campaign. As Carmody once said, "The Twilight Sleep is wonderful, but if you women want it, you have to fight for it."

The doctors in the German mountains were not the first ones to come up with the idea of painkillers during childbirth. A generation earlier, on January 19, 1847, Sir James Young Simpson, a Scottish obstetrician, gave ether to a woman in labor. He got the idea from William T. G. Morton, a Boston dentist, who had invited the curious to watch him use ether during a dental procedure at Massachusetts General Hospital. (MGH still calls the room "the ether dome.")*

---

* Jeff Mifflin, the archivist at Massachusetts General Hospital, told me that the ether dome was not called the ether dome in Morton's day. It had been called the operating room, then the old operating room, and not until the late nineteenth

Three months later, Fanny Appleton, the second wife of poet Henry Wadsworth Longfellow, became the first American woman to use ether during delivery. When every obstetrician in town refused to give in to her demands, she convinced her dentist, Nathan Keep, to bring some to her delivery. He brought a pipe for her to smoke while the midwife handled the birth. "All ended happily," Longfellow wrote in his diary. Baby Fanny was born with little discomfort to his wife. Six years later, Queen Victoria requested "that blessed" drug ether, ushering in a global phenomenon. Many women across the world took the queen's childbirth decision as a green light to go for the drugs. No longer were painkillers considered blasphemous or a mark of maternal weakness. (Yet there were plenty of doctors who thought that women who could not bear the discomfort of childbirth could not withstand the selflessness of motherhood.)

In addition to ether, they gave women opium, cocaine, quinine, nitrous oxide, and ergot. Not all at the same time. Most of the drugs did not dampen pain. They were used to speed labor, which in some ways is pain relief if it gets the pain over with faster.

The notion of twilight sleep emerged from pseudoscientific hogwash that the biochemistry of the modern middle-class woman made her highly susceptible to pain. According to this dubious theory—promoted by male doctors and female journalists—industrialized society corrupted the female body in such a way that she was too frail to withstand things that used to be natural. Put in simpler terms: modern science proved that women could not hack childbirth. Dr. Kronig, the twilight-sleep inventor,

---

century or early twentieth century did it acquire its latest label. What's more, Mifflin said that Morton, famed for administering ether in the first public demonstration of surgical anesthesia on October 16th, 1846, had claimed to have a dental degree from a school in Baltimore, but no record of his degree has ever been found. Morton was never on staff at MGH.

said that natural childbirth is impossible for middle- and upper-class women: "The modern woman . . . responds to the stimulus of severe pain more rapidly with nervous exhaustion and paralysis of the will to carry the labor to a conclusion. The sensitiveness of those who carry on hard mental work is much greater than that of those who earn their living by manual labor. As a consequence of this nervous exhaustion, we see that precisely in the case of mothers of the better class the use of forceps has increased to an alarming extent, and this where there is no structural need for forceps. " (Ohio University historian Jacqueline Wolf has referred to this period as the era of the fashionably sick.)

Word spread that not only were drugs better, but natural child-birth was dangerous. An article in *McClure's* claimed that natural childbirth can drive women crazy. Anna Steese Richardson organized the so-called Better Babies Movement that encouraged women to use drugs on delivery. In their book about twilight sleep, authors Tracy and Boyd proclaimed that "not a few women of good normal minds have gone to seed, become dumb, patient, brooding animals after the exhaustion of a succession of painful births." Doctors echoed similar sentiments. In 1914, Dr. X. Rhongy said that "the advent of this artificial and strenuous age" made natural childbirth virtually impossible. He said that turn-of-the-century women had a "distinct pathology" because of their sedentary lifestyle, diet of rich foods, and steam-heated houses.*

This notion of a weak female physiology, ironically promoted by feminists, pertained to middle- and upper-class women only.

---

* In an article by Anna Steese Richardson, she quotes George W. Kosmak, editor of the bulletin of the Lying-In Hospital, who once said: "There was a time when pregnancy and labor were conducted under much more natural conditions than at the present; but the development of our civilization has manifested during the child-bearing act as much as they are in any other connection."

Poor women were capable of withstanding the pain, which was a good thing, because they could give birth and get back to work right away. No need to give the nanny maternity leave. She could be back within days to take care of your children while you lounged in the postpartum bed. Drug advocates convinced women that taking medicine was not a sign of weakness but was healthier for themselves and their newborn. Pain hindered the baby's journey through the birth canal. The bumpy ride from womb to world would damage the child forever.

The talk was that upper-class women were so scared to give birth that they were having fewer children or none whatsoever and the nation would be overwrought with the offspring of poor immigrants. As Dr. Woods Hutchinson put it, "It is obviously a discouraging outlook for the future of upward progress of the race if all women of intelligence and culture on that account deliberately draw themselves out of the race stream, leaving only women of the 'lower type' to become the mothers of the next generation." He pointed to the worrisome comment by Mrs. Robert Brucke Liggett, of Minnesota, who supposedly told him that birth is so painful for the middle class that "the propagation of life should be confined to women of the lower type who are not capable of suffering."

The twilight technique was adopted quickly by a small clique of in-the-know American women, thanks, in part, to upbeat reporting. The headlines alone were inducements:

"PAINLESS CHILDBIRTH"

"LIFTING THE CURSE OF EVE"

"TWILIGHT SLEEP IS NECESSITY, NOT LUXURY"

"DRUG BOON TO WOMEN: NEW TREATMENT FOR CHILDBIRTH
CALLED MEDICAL MERCY"

You could not have asked for a better public relations campaign. *McClure's* ran a misleading article titled "Painless Childbirth," but twilight sleep was anything but painless. The 1914 article was written by journalists Marguerite Tracy and Constance Leupp, who were sent to the Freiberg, Germany, clinic but never saw any births. Their story relied on interviews with obstetricians who were promoting the method and with women who had the drugs but did not remember anything. The article included a photograph of a naked 5-year-old. The caption ran: "Careful observation of the scopolamine-born children has shown that no after ill effects can be traced to this method." The writers said that twilight sleep was the most "humane" way to bring a baby into the world. With the help of inaccurate reporting that presented twilight births like a boozy night out, the German birthing method became one of the most talked about topics during the roaring 1920s.

What the cheerful stories omitted was perhaps more telling than what the wonderful testimonials included. The articles did not mention that twilight sleepers were blindfolded and restrained while they writhed and hollered during delivery. To keep women still so that doctors could catch the babies, nurses tethered their arms to the bed rails and their legs to the stirrups with leather straps. Some rooms had "cribs," padded, deep beds to prevent women from falling out when they were twisting about. Nor did the reporters mention that the childbirth drugs diminished uterine contractions and caused hallucinations. It also impaired newborn breathing. Doctors called it newborn oligopnea, halted breathing. The doctors reassuringly added that while some twilight babies stop breathing for a bit, they never turn blue.

Women were enamored. "A whole new day had dawned on the human race," one writer gushed. "Men looked into the eyes of women and were unashamed; and women looked into the eyes of men and were unafraid. The birth-curse was broken." Hazel Harris, a twilight sleeper, called it the "greatest boon to mother-

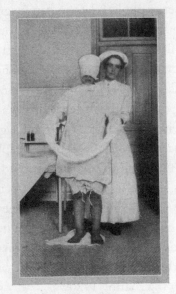

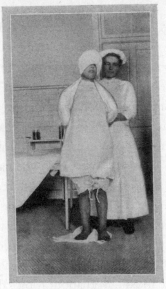

PLATE XIII.
GOWN WITH CONTINUOUS SLEEVE.

PLATE XIV.
GOWN WITH CONTINUOUS SLEEVE BEHIND NECK.

Woman prepared for twilight sleep. Bertha Van Hoosen, *Scopolamine-Morphine Anaesthesia* (Chicago: House of Manz, 1915). Courtesy of the Malloch Rare Book Room, New York Academy of Medicine Library.

hood in the world." Doctors said that twilight sleep worked best on smart women due to the biochemistry of their nervous systems. The *New York Times* published a 26-line ode to twilight sleep beginning with a line from Wordsworth: "Birth is but a sleep and forgetting."

With her journalist's flare for hyperbole—and for finding connections between two seemingly unrelated events—Marguerite Tracy equated the national impact of twilight sleep to World War I. "The war—and Twilight Sleep were the two news features of the fall of 1914," she wrote. "The subjects stood out strikingly against each other in periodical literature: on the one hand the brutalizing and cheapening of human life by war, on the other the giving of painless birth: one representing the hopeless past; the other the hopeful future; one used authority by men; the other representing the repudiation of medical authority by women."

Carmody raved about her German birth escapade during her national speaking tour. She told women how her birth experience included several weeks to relax in the clinic before delivery and ten days for recuperation afterward. Ten days was considered a speedy postpartum recovery compared with the typical two-week stays back home. The difference was that unlike the typical hospitalized American woman who was bound to bed for weeks afterward, the German clinic patients were allowed to walk around the room by day three and ride in automobiles a week later. Word was that the new drugs energized and hastened recovery compared with women who were aware during labor.

The flurry of glowing newspaper stories vexed physicians who worried about side effects. The same drugs had been available in the United States for other procedures, but American doctors were reluctant to give them to women in labor. A Long Island doctor wrote a letter "hysterical women carry the day" to the editor of the *New York Times* calling its twilight-sleep stories an "unprofessional and dangerous way of misinforming the public." He insisted the newspaper tell readers that the drug cocktail "depresses high nerve centers, suppresses important physiological processes, and produces grave, if often transient changes in bodily organs." In other words, the drugs can kill.

It's no surprise that the more doctors said they feared the drugs, the more women interpreted their reluctance as callous and narrow-minded. It could not have helped that the doctors' comments sounded more patronizing than informative. And while this was the way doctors talked to patients then, it could only have served to increase the rift between the advocates and the skeptics, between the women and their doctors. Charles Meigs, a respected obstetrician, said that a woman should not decide for herself how she should give birth because she "has a head too small for intellect and just big enough for love." Dr. Albert S. Barnes said women were making foolish choices because during pregnancy they are not "in full possession of their normal reasoning powers . . . and in a state of nervous equilibrium." He said that doctors were paving the way for a time when "hysterical women carry the day."

Skepticism did not dissuade an eager public. The *American Journal of Clinical Medicine* warned doctors about confrontational clientele. The "astonishing fact is that it [painkillers] has been around for 10 years and only met with condemnation. Why? Simply because the laity has taken the matter into its own hands." The word among women was that doctors were not worried about side effects, but worried about spending hours with a woman in labor. Twilight sleep required constant medical attention to ensure that the correct amount of drugs was used.

The barrage of patient advocacy groups held sway over public opinion—at least for a time. Women philanthropists sponsored twilight-sleep maternity in Chicago, Boston, and New York. A few New Yorkers converted a beaux arts mansion in a tony residential neighborhood along Riverside Drive into a maternity hospital. Then they found a young doctor willing to give twilight medicine. Neighbors were enraged. They claimed that the screams of women in labor and ambulances carrying away the dead ones would keep them up at night. A follow-up article in *McClure's* declared that "pressure from the women themselves" convinced doctors to give birth their way. Feminists declared a victory.

"Although she is dead, I have lost none of the interest."

On August 21, 1915, Charlotte Carmody died during the birth of her fifth child. She was given the twilight-sleep drugs and then bled to death. Her husband told the *Washington Post* that the drugs had nothing to do with her death. "Mrs. Carmody and I have both been enthusiastic about twilight sleep," her lawyer-husband Francis was quoted as saying. "Although she is dead, I have lost none of the interest." Her death during childbirth did not bode well for the cause. The Twilight Sleep Association died shortly thereafter. Carmody's next-door neighbor, Alice Olson, launched an Anti–Twilight Sleep Association.

In truth, twilight sleep was killed by its own hype. Carmody's death did not help, but it was not just about her personal tragedy. When a woman's birth experience was not as glowing as journalists predicted, she felt cheated. Ever so slowly, women began to turn against the movement, a crusade that at one time seemed so wonderful, so perfect. Maybe accurate reporting would have been better in the long run, rather than what one journalist dubbed "enthusiastic hysteria." Shortly after the fad started, women started reporting horrible reactions. Johns Hopkins stopped using the drugs altogether.

The rise and eventual demise of the National Twilight Sleep Association begs one question: Why did feminists embrace a toxic technique that diminished a woman's role in the birthing room? The answer has to do with the growing distrust of doctors. Women were eager to try a birthing technique that most doctors shunned. The National Twilight Sleep Association emerged in New York City amid a cacophony of political activism. It came on the scene just when the field of obstetrics itself was suffering from growing pains. Physicians specializing in obstetrics, the vast majority male, were trying to bolster their credibility among patients and peers.

At the same time, feminist organizations were gaining a voice in national debates. Doctors, instead of joining forces with women, felt threatened. Women, in turn, perceived their doctors' skepticism as a direct challenge to female authority, not a fear of the science itself. The upshot? Patients fought for innovations regardless of the consequences. In this case, they demanded medicine that on some occasions had hazardous consequences and on every occasion diminished a woman's authority in the birthing room. Women were fighting for the right to be in a complete stupor, allowing their doctors to call all the shots during the delivery. Ironically, women considered the crusade for pain relief a campaign for equal rights. They may have gained a say in choice, but they certainly lost control.

There's another way to look at it. Feminist women in 1914 and 1915 did not believe they were relinquishing control, says historian Judith Walzer Leavitt in her book *Brought to Bed*. They were demanding the right to give birth the way they wanted to. And if they did not want to remember it, so be it. Maybe they had memories of the utter pain of it all. Maybe those memories made the entire pregnancy a nightmare. If they could just erase those few hours when the baby was coming out, they were free to enjoy the pregnancy and the first few hours of motherhood. Forgetting the moment of birth should not make any difference whatsoever on the kind of mothers these women would become. Epidurals, which remove pain but allow women to be alert, were a long way away.

Part
Three

7

What Was She Thinking?
Freud Meets Fertility

In the fall of 1941, a 35-year-old woman called Sylvia started psychoanalysis. There was nothing unusual about that. She was a New Yorker, and Freud was fashionable then. But something was different about Sylvia. She was not looking for happiness. She was looking for motherhood. After years of trying to get pregnant to no avail, she and her husband adopted a baby girl. But within months, a visiting social worker deemed Sylvia unfit for mothering, took the baby back, and presented this ultimatum: If Sylvia ever wanted to adopt again, she had to endure a stint of psychoanalysis. Sylvia headed for the couch.

Sylvia was referred to Dr. Edith Jacobson, a recent German immigrant who would one day become a prominent psychiatrist and the first female president of the American Psychoanalytic Society.* Jacobson had no experience in the psychiatry of

---

* Jacobson arrived in New York having escaped the Nazis. She had already written several scholarly papers on self-esteem and depression. In fact, she was so absorbed in her work that when she was imprisoned by the Gestapo for failing to

motherhood—or this newfound role of turning allegedly bad
mothers into good ones. Her interests were self-esteem and
depression. But as a newcomer to America,
Infertility is   she did not have her pick of the patients
a woman's   yet. It goes without saying that doctor and
"emotional   patient seemed an unlikely pair: a woman
boycott"   desperate for a baby who probably wanted
                        fertility drugs, not talk therapy, coupled with
a doctor specializing in depression who wanted patients yearning
for self-exploration, not a bigger family.

Actually, things turned out a lot better than either Sylvia or her
psychiatrist ever imagined. Sylvia got pregnant, and Jacobson felt
it was her doing—at least in part. Jacobson said Sylvia had suffered from psychogenic infertility. In medical jargon, that trans-
lates to infertility caused by one's psyche. In plain speak, it means
your thoughts made you sterile. Jacobson did not invent the idea,
but she did become a believer and advocate, thanks in part to
Sylvia—or, more accurately, Sylvia's pregnancy.

The thinking was that repressed fears and hostility derailed
brain chemistry. The commotion in the head disrupted hormones
that in turn withered ovaries and clogged fallopian tubes. What
were these women thinking? Freudian analysts discovered anti-
maternal thoughts lurking deep within the psyche, so deep that
the patients themselves did not realize how much they truly hated
motherhood. Infertile women who insisted that they desperately
wanted to be mothers were not in touch with their inner selves.
Samuel Siegler, an obstetrician-gynecologist at Brooklyn Women's
Hospital, put it this way: "The unconscious desire not to have a

---

reveal her patient list, she treated her fellow inmates and subsequently published
articles about their experiences.

baby may stimulate the ovaries to pathological growth, premature maturation of the follicles, discharge of ova not ready to
fertilize, and sterility." More simply stated, he called it a woman's
"emotional boycott." That is to say, it was a passive-aggressive
revenge on her husband. As if the inability to get pregnant were
not depressing enough, these women were told that they were
mad and vengeful to boot.

Psychogenic infertility was, to be sure, a theory without any
proof whatsoever, but one with seemingly logical explanation. It
unified exciting discoveries in physiology with the new ideas in
psychiatry. Emerging evidence had just shown that fear spurred
an adrenaline surge and made the heart beat faster—the so-called
fight-or-flight response. Extrapolating from that theory, it made
sense that lurking fears about motherhood could change the body
chemically. In this case, fear destroyed female hormones that in
turn quashed fertility. "It is well known that a person can learn
what has been called 'selective blushing,'
a process by which the chest or back may "So long as she
become red from embarrassment," wrote is controlled
two gynecologists in a 1951 *Fertility and Ste-* by her repro-
rility article. "Perhaps the pelvic congestion ductive glands,
might be called a form of blushing beneath
the collar." she will remain

In some ways, the talk therapy was a reac- basically the
tion against organotherapy, a health craze same loveable
promoting all sorts of wacky hormone and gracious
therapies. By the early 1900s, you were no homemaker."
longer in a bad mood. You were hormone
deprived. You were no longer infertile, you were low estrogen.
There were so many diseases to cure and to concoct. Women got
ovarian extracts for obesity, insanity, "psychic menorrhagia" (dan-

gerously heavy menstrual cycles), and "involutional melancholia" (menopause-induced depression). Morning sickness could be relieved with either a pinch of dried ovaries or daily injections of corpus luteum. (The corpus luteum is part of the ovary that makes progesterone during the second part of the menstrual cycle and prepares the uterine lining to implant the egg.) One doctor claimed he cured a woman's insanity by feeding her pieces of animal ovary. Unfortunately, she committed suicide before the therapy kicked in, but if she had not killed herself, no doubt she would have been a happier person thanks to the therapy. And that's not all. Doctors also used pituitary gland extracts and sheep placenta. Years later, a health magazine would tell women that hormones have "special appeal" because they "really have some of the powers which superstition has always ascribed to the sorcerer's potion."

In the meantime, serious scientists were unraveling the truth of female physiology. They isolated estrogen and quickly thereafter progesterone. The supply of hormones helped basic scientists detect even more clues about how our bodies worked. In 1923, when the field was in full swing, Dr. Edgar Allen, of Washington University of St. Louis, and Dr. Edward A. Doisy, a biochemist from Yale University, figured out that a sharp drop in estrogen sparked the menstrual cycle. Accidentally—as many great discoveries happen—gynecologists Selmar Ascheim and Bernhard Zondek discovered that pregnant women excreted exceedingly high levels of hormones in their urine. In one fell swoop, they discovered a crucial clue about the physiology of pregnancy, the secret to one of the most effective pregnancy tests to date, and found a steady flow of hormones for further studies and for drug making.

When doctors first figured out what the ovaries and testicles were spewing, they reckoned they had found the Holy Grail of human reproduction. But that was simply foreplay. Gonads respond to a higher calling from the pituitary that dangles at the base of the brain. The pituitary releases chemicals that keep practically every function of the body in check. (For centuries, doctors knew that the pituitary squirted something, but they thought it was snot; the name *pituitary* is Latin for "slime.") But the story did not end there. Scientists soon discovered that the pituitary works under the direction of the hypothalamus, command central.

The hypothalamus spits out chemicals that turn on and off the pituitary. Doctors pictured the following scenario: the hypothalamus gives the green light to the pituitary to release its own sets of hormones that trigger the ovary to get going. It's a tightly controlled system. When the body is loaded with hormones, a feedback system tells the brain to slow down—or completely halt production altogether.

In some ways, hormone therapy nudged puberty, childbirth, and all things reproductive away from being considered a natural state to one that needed constant medical monitoring. To paraphrase Nelly Oudshoorn, author of *Beyond the Natural Body: Archeology of Sex Hormones*, women of the 1800s may have commiserated about their homebound lot in life, but they were not talking hormonal swings or PMS until the following century.

The vast assortment of remedies arrived at a good time for gynecologists, who were eager to broaden their clientele, to include more than pregnant women. With new drugs to treat all kinds of female woes, they could start caring for pubescent girls and menopausal women too. They could cure diseases never even considered diseases before. There were menstrual irregularities (which meant, of course, defining normal menstrual cycles), and menopause (which meant redefining menopause as pathology). As Sheila Rothman put it in *The Pursuit of Perfection*, "Gynecologists helped legitimize ovarian therapies and, in turn, ovarian therapies helped legitimize gynecology." In other words, the drugs were not just for the women. They helped launch many a man's career.*

Doctors have always had a hunch that chemicals in the body controlled not only the menstrual cycle but women's moods. Now that they nailed the brain-ovary connection, they felt they had women figured out. In 1939, in the keynote address at the 1939 52nd annual meeting of the American Association of Obstetricians, Gynecologists and Abdominal Surgeons (now called the American College of Obstetricians and Gynecologists), association president James E. King told his male-dominated audience that it was time for gynecologists to take over the field of female mental illness, closely linked to hormones. "There is no subject which should have greater interest for this Association than a discourse of woman herself," he announced at the meeting. "Not a discussion of her beauty or her diseases, for this we know, but rather an attempt by fact and fancy for her peculiarities and to explain her inconsistencies, and those delightful surprises we so often experience in our contact with them."

Dr. King concluded: "Nineteen years ago I presented a paper

---

* Men were not immune from the health fads. In 1889, 72-year-old Charles Eduoard Brown-Sequard performed a very unscientific study—some would say an outright silly stunt. Brown-Sequard injected himself with pieces of guinea pig and dog testicles. He told the world he never felt better. His brains and bowels worked better than they had in years. Brown-Sequard didn't bother with stodgy medical journals that fact-check and require approval by other scientists. He called journalists who loved the story. Unfortunately, his glory was as short-lived as the rest of his life. He died five years later. That put a damper on his claims of hormone rejuvenation. As a footnote to that episode, goat testicles became a 1920s fad for men seeking virility, thanks to John "the goat doctor" Brinkley, who became phenomenally rich peddling his wacky claims to celebrities. And there were men sipping monkey sperm cocktails who thought they were getting a little pick-me-up elixir.

before this Association in an attempt to account for the physical and mental differences between man and woman. It was based on the then meager knowledge of the endocrine system." He said that by the end of the 1930s, doctors knew what made women tick. They knew that women were mere puppets controlled by hormonal swings. Gynecologists could diagnosis us and heal us better than any other professional. But he wondered, what would these new therapies mean for women? Would we become more fertile? More vivacious? Sharper? "One may wonder what her position will be in the next one hundred years," Dr. King said. "Will she, as some timid souls fear, mentally and physically dominate and enslave us as we in the past enslaved her? Probably not; so long as she is controlled by her reproductive glands, she will remain basically the same loveable and gracious homemaker."*

The lingering question, though, prompted by the flurry of new discoveries, would be: Do hormonal swings trigger mood swings or the other way around—do mood swings cause hormones to shift? The answers were not mere academic bluster but a crucial debate about the kinds of therapies women were offered. Do you give a woman psychotherapy or hormones? Does she need anything at all? Is she sick?

As Dr. Viola Bernard, a prominent New York City psychiatrist said at a Planned Parenthood meeting, drugs were like Band-Aids covering up the mess, but did not get to the root of the problem,

---

* Dr. King may be the illegitimate son of President Grover Cleveland. According to Gail Collins's *Scorpion Tongues* (New York: Harper Perennial, 1998), 73–84, gossip was that Cleveland impregnated Maria Halpin, a widow from Buffalo who had a baby boy that she named Oscar Folsom Cleveland. Grover Cleveland never acknowledged the child but he did pay for its welfare. Eventually, he had Halpin institutionalized and put the baby up for adoption. A Buffalo doctor adopted him and changed his name to James King, who grew up to be a prominent gynecologist.

the psychic disruption that started the hormone malfunction in the first place.

From the get-go, Sylvia's case intrigued Jacobson. She said that Sylvia's rail-thin body manifested male-mimicking behavior, a sign of hysteria and neuroses. Sylvia weighed about 100 pounds and boasted that she could shrink to 90 on a good day. Her long bony arms and legs were gray. "Definite glandular problems," Jacobson noted. "Lacked feminine roundings, her breasts were small, her legs, shoulders, and arms very skinny . . . slim boyish figure gratified her masculine wishes."

"A mature woman without children is the psychological equivalent of a man without a Male Organ"

Dr. Jacobson was aware that anorexic women stop menstruating. By the twentieth century, doctors knew that starvation garbled the messages from the brain to the ovary that kick-start menstruation. The Freudian twist was that it was not merely diet, but subconscious thoughts that altered brain chemistry. She looked for answers in Sylvia's childhood, dreams, and secret fantasies. When Sylvia talked about her past, Dr. Jacobson saw an infant weaned too early; a kindergartner with penis envy; a schoolgirl traumatized by her brother's circumcision; and an adolescent repulsed by her physical sexual development. The upshot was that Sylvia was so horrified with her own femininity that she craved the life of a man, and these deep-seated yearnings staunched her feminine hormones. Sylvia insisted that she was depressed because she could not get pregnant. Dr. Jacobson saw things the other way around. She said that Sylvia's lurking depression began when she was a baby, and her lifelong mental illness triggered the sterility.

Doctors do not come up with theories of disease in hermetically sealed laboratories. They are influenced by popular culture like the rest of us. In the 1940s and 1950s, women were supposed to be eager for motherhood and enthusiastic about running the household. They were expected to preserve the sanctity of the American family threatened first by the Depression and then again by World War II. Hollywood reinforced and helped to shape America's attitudes. *Father Knows Best* and *Leave it to Beaver* made every mother feel inadequate—not to mention how inept the childless woman felt. Women who did not fit the mold were considered mentally off. "A mature woman without children is the psychological equivalent of a man without a Male Organ," wrote Dr. Marynia F. Farnham and Ferdinand Lundberg in their 1947 book, *Modern Woman: The Lost Sex.*

Consumer magazines promoted the psychogenic infertility theory with stories about infertile women who got pregnant after they quit work and learned to truly enjoy homemaking, which, of course, changed their brain chemistry back to normal. A *Time* magazine article entitled "Sterility and Neurotics" quoted a doctor who told the 1952 American Society for the Study of Sterility that infertility may be "nature's first line of defense against the union of potentially defective germ plasma." In other words, women who cannot get pregnant naturally, should not.

Pageant told readers about Betty, a tomboy who married a submissive man. She did not get pregnant—until that is, they learned to play their normal roles in society. After the war, Betty's husband returned home a man (more assertive) and insisted they adopt a child. The baby turned Betty into a real woman (she started to like housecleaning and all things maternal), and sure enough, she became pregnant. "As the partner in the marriage who wore the pants, Betty had had an unconscious opposition to child-rearing," *Pageant* explained. "The influence of these psychic factors so affected her endocrine makeup that she was unable to conceive. But when she became a real wife and mother, if only by proxy, her endocrine secretions straightened out." Stories about women like Betty who got pregnant right after adopting encouraged doctors

to search for a scientific basis for what was considered a trend.
Bernard and others thought that somehow becoming a mother
to an adopted baby reprogrammed women chemically. Or as
Bernard explained at a medical meeting, adoption led to a "re-
regulating of neuro and endocrine activity, a sort of reversal of
inhibition" (an inhibition of motherhood, she meant). Years later,
statistical analyses would debunk the adoption-fertility link despite
what seemed like an abundance of anecdotes. An NIH study in
the mid-1960s compared 249 couples with unexplained infertility who adopted children with 113 infertile couples who did not
adopt. Some 35 percent of couples who were infertile and did not
adopt got pregnant without drugs compared with 26 percent of
couples after adoption.

For the most part, doctors did not believe that the psyche was
the primary cause of infertility, but it was listed—sometimes
last—as one of the reasons. Articles that sounded like women's
magazine material were sprinkled throughout respected medical
journals. One article included an anecdote

"What the
clinicians failed
to accomplish
analysis
achieved."

about a married female lawyer who finally
became pregnant when she switched to
part-time work. The doctor had this to say
about her: "As her attitudes towards herself changed, her pelvic physiology under-
went change and pregnancy then became a
delightful anticipation rather than a hateful obligation."

There were scattered tentative findings and an odd array of
investigations. One researcher diagnosed 25 of 29 infertile patients
with mental illness. The investigator was a proponent of the talk-
therapy cure and conducted his own analyses rather than hire an
outside observer. You can't help wonder whether he would be
more likely to find neuroses to defend his theory. The researchers did not say which came first, the depression/neuroses or the

infertility, but it was assumed it was a brain-to-vagina route. A British study of 1,000 women suggested that stress can clog fallopian tubes.

In another study that certainly reflected a belief in the power of mind over matter, Dr. Ruth Molton, a psychiatrist at Columbia College of Physicians and Surgeons, convinced a group of women to provide daily vaginal secretions and maintain a dream journal that included the day of the menstrual cycle. Her aim was to examine how suppressed thoughts (the kind that emerge in dreams) affected hormones.

Some doctors believed in a "threshold" theory of fertility. Every couple has a quantifiable level of fertility: a sum of biochemical, hormonal, and emotional components. One dangerously low or even slightly off-kilter variable nudged the couple toward infertility. This theory explained why a neurotic woman could become pregnant with medical treatment only, which one doctor called a "hollow triumph." Drugs skyrocketed fertility levels, outweighing the psychic factor. "A vital practical consequence of the multifactor causation is that even a minor shift in any single one may tip the balance for or against pregnancy," said Bernard.

Sylvia, according to her doctor, satisfied her unconscious desire to avoid motherhood. Her hatred of all things maternal sparked a chemical reaction that clogged her fallopian tubes and triggered spasms in her vagina. After several sessions of psychoanalysis, Sylvia started to change emotionally and physically (she developed "feminine contours") and got pregnant. "Her breasts grew and her inverted nipples became normal. After an uneventful pregnancy, she delivered a healthy baby girl, who is now three years old," Jacobson wrote. "What the clinicians failed to accomplish analysis achieved."

As for Sylvia, she, like so many other women, quit analysis despite her doctor's fervent belief that she continue for the good of her reproductive health and for the good of her children. Sylvia got pregnant two more times after her first child, but she aborted both pregnancies. Then she starved herself again, stopped menstruating, and declared herself happy—free of periods and free to follow her pursuits. She became a freelance journalist specializing in child psychology.

"Despite the undeniable therapeutic success, Jacobson concluded, "it appeared that her neurosis had the last word. That the therapeutic success was due to analysis cannot, in my opinion, be doubted."

The talk-therapy approach to fertility faded with the rise of the quick-fix fertility treatments, such as in vitro fertilization and powerful fertility drugs. The revised 1974 brochure of Resolve, a fertility advocacy group, told women that their psyche is not to blame. In other words, get off the couch and start demanding medicine.

"If you are not in harmony with yourself and your culture, you are stressed. That is not totally different from Freud."

But like all medical fashions that ebb and flow, the psyche and fertility is percolating once again through the literature. There are no hard studies, but it seems that like women of the 1940s, there are women in the early decades of the twenty-first century who are not completely satisfied with modern fertility treatments. To be sure, they are seeking help in larger numbers than ever before, but they are looking elsewhere, too, for cures or something to boost the odds of success.

Recent studies are lending a bit of credence to the mind-body concepts, prompted by Freud but with a more palatable terminology. Nowadays we talk about stress—not male-emulating repressed desires. Dr. Alice Domar, author of *Conquering Fertility: Dr. Alice Domar's Mind/Body Guide to Enhancing Fertility and Coping with Infertility*, said that her 1990 study found that stress reduction with behavior therapy helped 34 of 54 women in a fertility clinic get pregnant. "Depression," she said, "may reduce egg quality, delay the release of eggs, prevent the implantation of a fertilized egg, or decrease levels of hormones necessary for a fertilized egg to grow and thrive." Dr. Sarah Berga, chair of the Department of Obstetrics and Gynecology at Emory University, said that one of her studies that included 184 women found that those who received emotional support during infertility treatment were more likely to have babies than those who did not. "Having hypothalamic amenorrhea is more than a reproductive condition. When the hypothalamus is stressed, other things are stressed. Cortisol goes up, the thyroid goes down, and the fetus may not respond well to those problems," said Berga. In another small, yet scrupulous study, Berga showed that cognitive behavioral therapy alone restored ovulation in seven of eight women compared with two of eight women who did not get therapy. "The truth is that in an anthropologic context, if you are not in harmony with yourself and your culture, you are stressed. That is not totally different from Freud."

What was Sylvia doing to her brain chemistry that shriveled her eggs? Were her thoughts diluting the powers of estrogen? There was no proof. No one had ever watched a burst of hormones shrink ovaries or squeeze a fallopian tube. But like many scientific ideas, there was some sound reasoning, and it made sense at the time. It caught on for decades because it worked for everyone. It gave desperate couples a reason for their infertility. It goes without saying that talk therapy was a boon to the fertility industry, suffering from dismal results. In the 1940s, only one in four women became pregnant after a dubious assortment of treatments that included varying doses of thyroid and estrogen extracts and X-rays of the ovaries.

Psychogenic infertility was also a theory that suited gynecologists because they needed an explanation for the bulk of their patients in the "unexplained" category. In an odd sort of way, the new psychiatric outlook made women feel like they had more control over the process, inasmuch as you can control your emotions. If you could tap into your inner psyche, you could heal your dysfunctional self. As David Rosner, a professor of history at Columbia University, sees it, "Psychoanalysis was a liberating force." While pharmaceuticals were exploding and medicine was becoming more scientific, psychoanalysis, says Rosner, was seen as its humanizing sibling, the one field that considered each patient an individual with unique needs. If you look at it that way, women were not being told they were crazy; they were being offered a more tailored treatment that they would not get at the standard fertility clinics. Sure, the doctors were blaming the victim, but they were also handing over the controls.*

The gestalt of the 1940s may have planted the seeds for the first American natural childbirth crusade. Its founder, Elisabeth Bing, was a German immigrant and physical therapist for post-partum women who also struggled to get pregnant. She had been referred to a psychiatrist but found him judgmental and never went again. (She eventually got pregnant after being told to drink more orange juice.) Bing, like so many others, may not have bought into the talk-therapy cure for fertility, but she was not dismissive of the burgeoning mind-body links popularized by Freud and his disciples. Her focus was birth, not conception. Fitting for the time, she called her organization the American Society for Psychoprophylaxis in Obstetrics. That mouthful of a title would later be changed to the simpler Lamaze International.

---

* In a 1945 article in *Human Fertility*, Dr. Edward Weiss explained that the new field of psychosomatic medicine was "founded on the important advances in physical medicine as well as on the biologically oriented psychology of Freud." E. Weiss, "Psychosomatic Problems in Fertility," *Human Fertility* 10, no. 3 (September 1945): 74.

8

It's Only Natural

Elisabeth Bing, the founder of Lamaze International, said that some time in the late 1950s, she convinced her friend Marjorie Karmel to smuggle an illicit French film into America. Karmel was on vacation in Paris and slipped the contraband into her luggage. The following week, the two women organized a private screening at Karmel's Upper East Side apartment.

The movie was an explicit portrayal of a woman having a baby naturally. *Naissance* (French for "birth") was shown routinely in France to expectant parents. The point was to take the mystery out of the birthing process and in turn allay fears. (The women in the film did not seem horribly distressed.) But New York was a different scene altogether. Here *Naissance* was deemed obscene. Bing said that the 92nd Street Y in Manhattan turned down repeated requests to include it in prenatal classes. It would take a year for her to convince Mt. Sinai Medical Center to show it.*

* This was not the first time a birth movie riled Americans. In 1938, *The Birth of a Baby* was made with what *Life* magazine called "the high purpose . . . to reduce

Birth for most mid-century women was anything but natural.
As historian Jacqueline Wolf tells it in *Delivery Me from Pain: Anesthesia in Birth in America*, women were getting all kinds of anti-psychotics and other medicines to knock

"Happy child-
birth is the
most vital fac-
tor for building
a progressive,
purposeful and
considerate
world."

them out, including Thorazine, Benzedrine,
nitrous oxide, and Demerol. But Bing and a
small group of trailblazers were determined
to reverse the trend that began with twilight
sleep. Following a path already cleared by
natural childbirth gurus in Europe, they
started to pave a new American way of
thinking about reproduction. The early cru-
saders were not rebelling against drugs or
doctors. Not yet. They preached psychology

over pharmacology, much the way the Freudian analysts tried to
cure infertility. They taught mind-body tricks to replace the urge
for anesthesia. They coached pregnant women to breathe deeply
and rhythmically, part of a standardized program touted to train
the body to relax automatically. In essence, natural childbirth pre-
1970 was not a revolt against anything. It was a natural progression
of the burgeoning field of mind-body research, or as the *Washington Post* remarked in 1949, "an advance along the lines of psychosomatic medicine." Doctors investigated how the psyche spurred
asthma, ulcers, rashes, and so on. They used the tools of Freudian
psychoanalysis to diagnose infertility. Why should childbirth be
exempt from the paradigm?

Believers said that natural childbirth eased the trauma of deliv-

---

illness and mortality among mothers and their babies." The movie was made by
the American Committee on Maternal Welfare. Local doctors had to sanction the
film before it could be shown to the public. Unlike *Naissance*, *The Birth of a Baby*
shows the newborn emerging through drapes of sheets.

ery for the baby. A mother with a relaxed mental state oozed those feelings into her newborn baby, planting the right seeds for the child's emotional well-being. Psychiatrists believed that a baby who entered the world snuggling with an alert mother had a head start, emotionally speaking, compared with a newborn swept from birth canal to nursery with mother in a drug-induced stupor. Internationally renowned psychoanalyst Helene Deutch urged women to give birth naturally for the psychological sake of the offspring. Birthing guru Dr. Grantly Dick-Read once effused: "Happy childbirth is the most vital factor for building a progressive, purposeful and considerate world. Let us help them to realize birth should be natural . . . and we will have healthier children with controlled nervous and mental function."

Incorporating popular Freudian views, some natural childbirth advocates warned that unconscious fears materialize during labor and delivery and affect a mother's perception of childbirth, which in turn influences the mental state of the newborn. In other words, they believed that women who felt pain during birth were manifesting lurking concerns about their own childhood or their relationship with their mother or their love for their unborn baby. If you were the sort of woman, say, who had become exceedingly attached to the baby in the womb "as if it were another limb," the delivery would be as scary as an amputation. Or let's say you had trouble toilet training because you believed you would lose control; labor would evoke those horrible out-of-control feelings.

While each birthing guru preached a unique variant of natural childbirth, the underlying premise was the same: anxiety tenses muscles, and tight muscles increase pain. Some preachers taught women to relax muscles through exercise, some through meditation, some through religion. A Beverly Hills gynecologist promoted hypnosis, and another doctor promoted an exercise to relax the cervix and vagina and told women to practice during routine pelvic exams. "A tense woman," Dick-Read explained, "is closing the door against her baby."

The pundits also said that the way a woman coped with her pregnancy and delivery predicted the kind of mother she would become. A mother who was awake during delivery and held her newborn immediately afterward would raise a normal, nurturing family. If you could not cope with childbirth, maybe you could not hack motherhood. Such notions about childbirth were bantered about in medical journals, which does not mean they were widely accepted, but it does indicate that they were parlayed among childbirth experts.

In some respects, medical advances fueled the natural childbirth movement. Women were no longer grateful just to survive; they wanted to enhance the experience. Bing and the other pioneering women, those who asked their doctors for natural childbirth, were the ones who could either afford to go to Europe and have their babies delivered by Lamaze himself or the ones gutsy enough to demand that their doctors deliver their babies according to their wishes. Marjorie Karmel had her baby in Paris, much the way Charlotte Carmody went to Germany for twilight sleep.

Before the Americans entered the scene, two French obstetricians, Ferdinand Lamaze and Pierre Vellay, popularized a breathing technique touted to ease labor pains and then made the movie *Naissance* to prove that it worked. Lamaze did not call his technique Lamaze. Only the Americans did. It was not even his idea. In fact, Lamaze had nothing to do with Bing, despite the fact that she made his name part of the American vernacular. (He ignored her and Karmel during his one lunch with them and directed all his comments to the male obstetrician at the table. Karmel adored Lamaze despite his

"I got everything I raged against"

disinterest in the ladies, and she even wrote a book called *Thank You Dr. Lamaze.*)

Lamaze adopted the concept from Russian obstetricians. They got the idea from Nobel laureate Ivan Pavlov, who made dogs salivate at the sound of a bell.* They reckoned that if bells got dogs' juices flowing, there must be a similar mind game to make laboring women relax their womb. The idea spread rapidly around the world. A Japanese obstetrician introduced natural childbirth in his country after hearing about the Russian methods while detained in China during World War II. In deference to the Communists and with all intentions to snub Americans who were popularizing a similar method, the Japanese doctor called it "the Soviet-Russian and Communistic Chinese Method of Analgesic Delivery." He told a colleague in the United States that he taught the American style of natural birth, but "six years of American occupation made us tired of the U.S. and of American methods in many respects," so he went for a Russian-sounding name. The Communist nickname did not sell well, so the Japanese eventually went with "psychoprophylactic method of analgesic delivery."

Bing loathed the name "natural childbirth," preferring the less-headline-grabbing but more meaningful sobriquet "prepared childbirth." "Natural" sounded like a whole new approach, whereas "prepared" sounded as if they were simply informing you. She also opted for "prepared" over "natural" because she claimed that her goal was not to eschew drugs altogether but to empower women to make informed decisions, though her litera-

---

* Pavlov did not win the Nobel Prize for his famous salivation experiments. He got the Nobel in 1904 for his work on the physiology of digestion and then did the drooling tricks after. George Johnson tells the delightful tale of Pavlov and his dogs in Chapter 9 of *The Ten Most Beautiful Experiments* (New York: Alfred A. Knopf, 2008).

ture would suggest otherwise. Her mantra was "Awake and alert."
Natural childbirth teachers, Bing claimed, wanted to enlighten
expectant parents, who for too many years had been in the dark
about childbirth, both figuratively and literally.

There is one little secret that Bing—Ms. Awake-and-Alert—
does not often share, particularly with her devoted disciples. After
years of preaching prepared childbirth and empowering maternity
patients, the mother of the movement gave birth as unnaturally
as can be. "I got everything I raged against," Bing admitted about
a half-century postpartum. It's easy to imagine that Lamaze students who could no longer stand the pain and begged for a whiff
or a shot of anything felt pangs of guilt afterward. And yet, their
guru seemed to acquiesce without any qualms whatsoever. "I had
the works," said Bing with a grin. She said she was sedated with
laughing gas and got an epidural all because she pestered her doctor. From the little she remembers of her son's birth, she did recall
thinking the doctors were having trouble getting him out, so she
kept asking, "Is my baby all right? Is my baby all right?" Finally
her doctor said he could not concentrate with all the chatter so he
gave her drugs and told her she'd be a better birth teacher for it.

"Woman is made primarily in order that children might be born into this world."

"I had watched enough births to see this can happen," said Bing. "I just didn't anticipate it for myself."

Still, for all the talk about so-called pre-pared childbirth or fearless childbirth or Lamaze or whatever the moniker, few Americans outside of this tightly knit circle were talking about it until the whirlwind
book tour of Dr. Grantly Dick-Read's in 1947. And then what had
been banter among medical circles became headlines in national
newspapers.

The Maternity Center Association, a New York City–based

organization that promoted birthing education, sponsored Dick-Read's American tour and launched a huge marketing campaign to ensure its success. They bombarded the press with his books and press releases. "The Association wanted to make it clear that here was no second Coué [a turn-of-the-century psychologist who promoted the power of positive thinking] but a distinguished man of science with penetrating ideas, who had the interest and backing of many important figures in obstetrics and was invited to the United States by the Maternity Center Association," said one internal memo. His first lecture, on the evening of January 19th, 1947, was well attended by 369 American obstetricians. Another 360 professionals showed up at a second talk at Lenox Hill Hospital in New York City.

The timing could not have been better. He arrived in America when midwives, mothers, and even a few physicians were eager for a change. *Childbirth without Fear* hit a nerve among the small yet vocal group of disgruntled women, who considered standard births overly intrusive, overly medicalized, and downright dangerous. Midwives were angry because they were losing turf to the obstetricians. Unlike the English, who began to train nurses for midwifery, American doctors tried to get them off the playing field altogether.

Still, in many respects, Dick-Read seems an unlikely father of natural childbirth. He was an egotistical male chauvinist with no tolerance for skeptics. He was religious and believed that childbirth should be a heavenly affair. He was not a board-certified obstetrician, yet he sired a global phenomenon. If nothing else, the man was a marketing wizard. His 1942 *Childbirth without Fear* was a best seller. He put his name on maternity lingerie, peddling a bra attached to humungous underpants with a complicated series of suspenders. He claimed that his underwear supported the growing belly and breasts. It looked more contraption than lingerie.

Just the way it had always been for him at home in England, he was adored by the media and by women but eyed suspiciously by colleagues. (In England, there was so much rancor between him and his colleagues that he moved to South Africa, where he started his own maternity clinic. He also left his wife and children back in England and married his nurse, who was much more of a cheerleader for his cause.)

A devout Christian, Dick-Read preached that the moment of birth should be a divine experience. His evangelical approach boosted his popularity during the religious revival of 1950s America. Christian organizations pushed natural childbirth. One religious Dick-Read devotee published a popular *Natural Childbirth and the Christian Family*. In 1956, Dick-Read was granted an audience with the pope and awarded a Silver Papal Medal.

To this day, he is considered a hero among feminists, though he preached that "normal" women only found true happiness in the home. In his first public speaking engagement in America, he said that God made men and women different intentionally. "Woman is made primarily in order that children might be born into this world . . . yet somehow we don't always feel that women are quite as proud of that magnificent gift as they should be." He believed that a pain-free birth is how God intended birth should be. Or as he put it:

It may be as a child you were interested in your dolls, your needlework, your games; at puberty your mind widened, and you found beauty worked out for yourselves. The mysteries of those startling strong emotions which swept you from the age of twelve or thirteen . . . maybe to twenty. . . . Then in fulfillness of time it became your lot, and you met real love, and with that natural urge progression subtly unrecog-

nized by you except in terms of ambition and beauty, you
acquired and became matured, and so to marriage, and then
to pregnancy, and if healthy-minded carefree and happy you
watched, felt and dreamed of your developing child. The full-
ness of your ambition became clear to every one of you. You
felt some instinctive closeness with the Creator. . . . Then the
kindly gods which had led you through all this turned around
and said, "For that, I will give you hell. What nonsense!"

Dick-Read believed that perceptions influenced reaction. In
other words, if you think childbirth is scary, you will tense your
body and realize your fears. What irked colleagues was not so
much *what* he argued—his ideas made sense to them—but *how* he
argued. Dick-Read wrote as if he were the only gynecologist to
understand the mind-body connection and the only doctor who
cared about patients. There were other gynecologists promoting
similar ideas; they just did not have his sales prowess or the Dick-
Read way with women.

His followers considered him the epitome of the caring, kind
doctor. But Dr. Dick-Read complained bitterly about neurotic
middle-class women who worked themselves into a tizzy over
labor and delivery. Labor pain was a female delusion, he said. It
irked him that animals gave birth without pain and yet humans
whined. (He did not give any credence to physical differences:
the pelvis among four-legged animals is wider than the human
pelvis.) He said that spirituality and meditation should reeducate
upper-class women to deliver babies the same way poor people
do, without pain.

Nothing in Dick-Read's book is more disturbing than the blame
he places on women themselves for conjuring up a pain that he
claimed did not really exist. "From this inestimable gift [of child-

birth] emerges the power of mother love, which forms the pattern of the infant's psyche as surely as mother's milk fashions its physique," he wrote. Psychoneurotic women, he said, suffered during labor. Normal women gave birth easily. If one of his patients experienced a difficult birth, it was her fault for being a "selfish, introverted woman."

Dick-Read's clientele ignored the pejoratives and saw an obstetrician willing to listen to women and help them overcome fears. In a way, they were right. Dick-Read tried to write personal responses to every woman who wrote to him. There were boxes of fan mail. *Post-War Mothers*, a book by historian Mary Thomas, contains 64 of the 3,400 letters between Dr. Grantly Dick-Read and his patients between 1933 and 1959.

October 26, 1955, from Kentucky: "Dr. Read, your book is truth, and I cannot understand why every woman and doctor will not open their eyes to this truth."

April 14, 1947, from a medical student in Illinois: "I wanted to write you to tell you how enthusiastic I am about your progress, and how much I hope you will be able to continue your work and train students along the lines of your theories."

January 18, 1952, from a mother in Washington: "I wish I could express the gratitude I feel for the new confidence and clarity of vision I find are mine due directly to your efforts."

When Dick-Read came to America, one of the first centers he visited outside of New York's Maternity Center was the obstetrics department at Yale University. Yale would become one of the first U.S. centers to launch a natural childbirth ward, thanks in part to a

Grantly Dick-Read's American tour in 1947. Courtesy of the estate of Grantly
Dick-Read, Pollinger Ltd., and the Wellcome Library, London.

fortuitous confluence of eager nurses, obstetricians, and pediatri-
cians swayed by Dick-Read's dogma. It blossomed because of the
young mothers themselves who demanded kinder, gentler births.

The mothers called themselves the Yale Dames. They were a

loosely organized group of medical-student wives known more for baking cookies to serve between afternoon classes than for stirring up controversy. These feisty women, not the doctors,

"It's my hope your story will . . . convince apathetic medical people that they are going to have to start being a little more cooperative."

were the ones who first learned about Bing and Dick-Read and demanded that the hospital and the doctors at Yale change their ways. Their confidence to question medical authority probably had to do with their age and close alliance to the profession. Many of them were married to former World War II GIs who started medical school after their stint overseas, so they were older than the typical medical student and closer in age to some of the attending physicians. One former Dame told me that because they socialized with physicians, they did not have the typical fear of doctors. They spoke up. They demanded what they wanted.

These women were meeting regularly to talk about parenting and childbirth and recipes and what not, while their husbands were on the wards learning to be doctors. This seemingly innocent chitchat morphed into an inspiration for change. "Even before the sixties, we had discussions about controlling our own destinies," said Irmgard Wessel, a friend of the Dames and wife of a Yale pediatrician. "I think the young doctors' wives all felt natural childbirth was the way to go," she said. Wessel and the ladies also pushed for rooming-in, the notion of having the baby stay in a bassinet with the mother rather than down the hall in a nursery. Her husband, Dr. Morris Wessel, would become a world-renowned rooming-in expert and spokesman, ideas he admitted were sparked by the Yale Dames.

Priscilla Norton remembered those days well. She was one of

the very first patients on the new maternity ward at Yale. "I had /
a doctor for a husband and a best friend who was an obstetrician
and neither one of them did anything to prepare me for childbirth," for the birth of her first child. "All my mother kept telling
me [about delivery] was 'it's so awful, it's just dreadful.'" Afterward, she met the Yale Dames and remembers them talking about
how little they knew about childbirth and wondering whether
there was a better way. Five years later, Norton attended Yale's
natural childbirth classes and then gave birth to her second son
in the rooming-in section of Grace New Haven Hospital without
any anesthesia at all.

The women, of course, could not have made headway without the help of willing physicians and nurses. Dr. Edith Jackson,
for one, was a psychoanalytically trained pediatrician who would
spearhead the rooming-in concept. Dr. Herbert Thoms was a
young obstetrician at the time who would become a national advocate of natural childbirth and a Dick-Read disciple. He worked
closely with his mentor, Dr. Frederick Goodrich Jr., to launch a
prepared-childbirth center and to publish scientific studies on its
success. One study found that 19.3 percent of 156 women who
were coached for natural childbirth did not use any drugs at all.

The Yale team started prenatal classes to teach women about
labor and relaxation techniques, à la the Dick-Read style minus the
religion. The most dramatic change was the reorganization of part
of the maternity ward—a colossal undertaking. The new interior
design allowed for groups of four mothers and their babies to
room together, a system that encouraged maternal bonding with
babies and with each other. The new decor was inspired by Dr.
Jackson, who did a stint in Vienna studying Freudian psychoanalysis. Yale not only nurtured the rooming-in culture, but studied the
psychological benefits to both mother and baby, following them
for years after. The Yale team pointed to scientific studies show-

Women in a Grantly Dick-Read prenatal class. Courtesy of the estate of Grantly Dick-Read, Pollinger Ltd., and the Wellcome Library, London.

ing that profound psychiatric dangers lurked within mothers and babies when mom slept through the birth and baby was whisked away to the nursery. They called the standard nurseries, where babies were clustered in glass-enclosed rooms, "an unnatural fragmentation of the family." Other hospitals followed suit, including George Washington University Hospital in Washington, D.C., and Los Alamos Hospital.

As Dr. Wessel saw it, the popularity of psychoanalysis gave the natural childbirth and rooming-in crusades a much-needed boost. Babies needed to start as close as possible to mom for the mental well-being of both mother and child. But it was money, pure and simple, that made the campaign take off. Wealthy women wanted natural childbirth, and they wanted to room with their babies. If their own doctors did not provide the care they wanted, they went elsewhere. Savvy doctors, even the most cynical, realized that they needed to offer prepared childbirth to keep their patients happy. The Yale doctors conceded that hard proof was lacking, but that wasn't as important as their own experience. "Whether this assumption is scientifically correct from the viewpoints of anatomy, physiology, and psychology is not too important," Drs. Goodrich and Thoms wrote. "Our experience has taught us that it . . . works."

The press was all over this story. Women's magazines and newspapers were filled with glowing stories about the natural methods of birthing. "Motherhood without Misery," wrote *Collier's*; "Cruelty on the Maternity Ward," wrote *Ladies Home Journal* about hospitals that refused natural childbirth techniques; "I Had My Baby without Pain," declared *Baby Talk* magazine.

Testimonials abounded: "I was alive and elated," wrote one woman. "In less than a minute of concentration, the pain was completely gone," wrote another. *Childbirth without Fear*, borrowing Dick-Read's book title, was the *Life* magazine cover story in January 1950. The photo essay showed a blissful mom with baby at her breast. The woman was not hypnotized, not drugged, and not completely pain-free either. She was without fear, as the title said. (There were debates in women's magazines whether the point of natural childbirth was to remove the pain, remove the fear, or both.) *Life* included head shots of Mrs. Charles Barnes as her labor progressed. There's an early shot of her relaxed and playing solitaire, an action shot with her squinting, and the postpartum photo of a joyous mom.

In response to the spate of stories, women wrote letters to editors of local newspapers and magazines complaining that they could not find doctors willing to do childbirth the Yale or Bing way. After a sanguine article about the joys of natural childbirth, a reader responded: "It's my hope your story will . . . convince apathetic medical people that they are going to have to start being a little more cooperative." Another letter writer dismissed doctors for caring more about germs than patients. "Hospitals have concentrated so long on killing germs and keeping them away from patients, they have forgotten people. The pendulum is swinging back to the warm, friendly atmosphere of the late 1800s. . . . Dr. Read's success needs all the publicity it deserves."

Despite Dick-Read's media blitz in 1947, most of America was not quite ready for him. New mothers were reading Dick-Read's polemic and amused by the headlines, but for the most part they were not buying into it.

Natural birth is "anti-natural."

For most of the pregnant world, childbirth in the middle of the twentieth century was not an event to remember and certainly was not something to enjoy on the big screen—watching natural childbirth flicks, for instance. Maternal mortality dropped 70 percent from 1935 to 1948, and newborn deaths dropped by 40 percent. Most women had drugs and afterward stayed in bed, sometimes lying completely still for days so their bodies could heal. Bernice Levine, who gave birth in 1944 in Brooklyn was told not to "dangle until the eighth day." "That's what they called it," she said. "Dangling. You got to hang your feet over the bed." Few women complained because they were grateful to doctors for keeping them and their babies alive.

American birth was a means to an end. You went to the hospital, left your husband at the door, and awoke with a clean baby in your arms, soon to be whisked away to the nursery. After delivery, your baby visited you at times dictated by the nursing staff. It was all very much the way Lucy Ricardo did it in the 1953 *I Love Lucy* episode. She tells her husband Ricky to go work at the Tropicana Club while she gets a taxi to the hospital. He dances with his nightclub girls and arrives at the hospital after the birth, peering through a glass screen at a lineup of white-starched nurses, trying to figure out which newborn is Little Ricky.*

In *Lying-In: A History of Childbirth in America,* historians Richard and Dorothy Wertz say that childbirth became like the modern factory, the "processing of a machine by machines and skilled technicians." From our twenty-first-century perspective, births in the *I Love Lucy* days seemed factory-like, but most Americans perceived it as perfectly normal, actually better than normal. Births were sterile, scientific, and organized. And that was perfectly fine.

Doctors gave commands and handed out drugs. Women obeyed. In those days, for instance, parturient women who had suffered from dangerously high blood pressure—called preeclampsia—during their first pregnancy were warned not to gain more than 15 pounds the next time. It was an arbitrary cutoff. To "help" women, some doctors handed out amphetamine-laced diet pills or diuretics. Doctors thought that keeping women thin would prevent bloating, which they thought triggered preeclampsia, which can sometimes lead to seizures, coma, and even death. Except for one thing. Excessive bloating is a sign of preeclampsia, not a cause.† These were the sorts of treatments that natural child-

---

* The famous pregnancy episode, which aired January 19, 1953, made television history. Never before had a pregnant belly, albeit one clad in a huge-bowed maternity dress, been shown on TV. Lucille Ball, who really was pregnant, insisted. The script was approved by a priest, rabbi, and minister, who gave the green light as long as they did not use the word, *pregnant.* Lucy was expectant.

† *Eclampsia* is Greek for "bursting forth." Apparently, that's what eclamptic women looked like thousands of years ago, so doctors gave them that label and it stuck. The full-blown deadly consequences rarely occur nowadays because doctors look for the triad of warning signs—high blood pressure, protein in the urine, and swelling—and they will give preventive medicines right away. In generations past, women were purged and bled to rid the body of whatever toxins were triggering seizures. No one knew what was going on, but they figured a thorough housecleaning couldn't hurt. More often than not, the purging made women too fragile to survive. I. Loudon,

birthers would start to question. But the rest of the world was not ready for their scrutiny of medical authority.

Critics blamed the natural birthers for retreating to the days of yesteryear, when women suffered and often died. They thought that these women were denying themselves crucial medicine and lifesaving interventions. They cried negligence. They lambasted their colleagues for depriving patients of modern medical care. What's more, they said that telling pregnant women all the details of gestation increases anxiety. In a scientific publication, Arthur Mandy, a Baltimore-based obstetrician, wrote that prenatal classes, particularly a lesson with "too intensive a program of education in anatomy and physiology," is destined to "arouse as much anxiety as it allays." He criticized the Yale findings that showed that women who went to their prenatal classes needed fewer drugs than those who never attended. In response, Mandy pointed to a three-year study at Baltimore's Sinai Hospital that found that natural childbirth drove some women crazy. One of their patients, researchers reported, tried to commit suicide. Another became frigid. A third developed all-body pain and a rash, considered a psychiatric manifestation of being awake during birth. William Studdiford, head of obstetrics at Hartford Hospital, dittoed Mandy. He called the natural birth "anti-natural." A 1948 *Collier's* article showed an ecstatic Mrs. Bedard having a baby the Yale way, but according to a local New Haven newspaper, Mrs. Bedard later sued the magazine, claiming that she was not nearly as happy as the photos showed. The piece also sparked a flood of hate mail.

From Mrs. Wade Ellis, in Cleveland, Ohio: "Dear Editor: After reading Motherhood Without Misery (Nov 16th) I am thoroughly disgusted. . . . They call it pain caused by fear but let any man

---

Death in Childbirth: An International Study of Maternal Care and Maternal Mortality 1800–1950 (Oxford: Clarendon Press, 1992).

receive a slight injury and listen to the moans and groans. Is this pain also caused by imagination and fear?"

From Mrs. Jas P. Wilson in St. Louis, Missouri: "Dear Editor: So pain in childbirth is a vicious circle of the mind! If Dr. Sawyer really believes it, then he can have all the rest of my family for me. Frankly, I think he's nuts!"

Another letter writer called Dr. Sawyer "silly and nuts," and another asked whether any man would agree to have surgery without painkillers.

In the 1970s, natural childbirth would blossom into a celebrity-studded nationwide movement, an outgrowth of broader grassroutes campaigns to "demedicalize" childbirth. In its infancy, it was slow to develop. Bing and her coconspirators in Manhattan and New Haven did not make waves in the postwar years. Gently rocking the boat is a better analogy. More women than not who gave birth in the late 1940s and early 1950s don't even recall a natural childbirth brigade. But it was there. It was more a media event than a popular birthing method. Women who stayed awake and enjoyed the birthing experience became quotes in magazine articles. The movement would not sweep the nation for another 20 years. As Dorothy Thomashower, a New Yorker who gave birth unnaturally in the early 1950s put it, "Practically all of us got some anesthesia. It depended on how hysterical you were. It was based on the personality of the woman. It wasn't until women were burning their bras and letting their hair grow under their arms that they really all went for natural childbirth."

"It wasn't until women were burning their bras and letting their hair grow under their arms that they really all went for natural childbirth."

Why? For one thing, anger is a powerful motivator. Women in

the 1950s were not rebelling. If they wanted to tweak the medical routine—say, be coherent during delivery—they did so with their doctor's blessing. That would all change in the 1970s, when their daughters were marching for civil rights, joining feminist rallies, and demanding a patients bill of rights. Informed consent—telling patients everything that was about to happen to them—became dogma in the 1970s. Think about the *All in a Family* episode that aired on CBS December 2, 1975 (the same network that gave us the *I Love Lucy* episode 22 years earlier). Future TV grandparents Archie and Edith Bunker arrive at the hospital before their laboring daughter Gloria. When she does arrive, she is not nearly as coiffed as Lucy. Her hair is disheveled. She is tired and downright uncomfortable on the delivery table. Mike, her husband, is by her side, coaching her in the Lamaze way. She is awake and looking into a mirror so she can watch the baby being born.

Cut to Edith and Archie in the waiting room.

Edith: "Mike is coaching, he's helping Gloria feel better. I wish you'd been with me when I was having Gloria."

Archie: "I was with you when it counted."

There were women thinking like Gloria in the 1940s and 1950s, just not a lot of them—and they certainly weren't fictionalized on television sitcoms. In the 1940s and 1950s, delivery without drugs was an occasional request. The late Florence Wald, former dean of the Yale School of Nursing and founder of the hospice movement in the United States, saw similarities between the drive for hospice and the natural childbirth campaign, both of which she said were bolstered by 1970s health activism. Women, she said, finally learned that they could take back the control they once had about how they ended life and how they created it.

9

Toxic Advice and a Deadly Drug

DES

From 1938 to 1971, millions of pregnant women took high doses of a synthetic estrogen touted to prevent miscarriages. The scientific name for the drug was diethylstilbestrol. Most people remember it as DES. The tragic irony is that the drug did not prevent miscarriages but it did harm babies exposed to it in an insidious way. The dangers did not emerge for years after exposure. At its worst, DES triggered a rare and deadly form of vaginal cancer in about one in every 1,000 women whose mothers took the drug. The cancer struck when women reached their teens and 20s. Some DES cancer survivors have had their vagina, uterus, and fallopian tubes removed.

Thousands, perhaps millions, of so-called DES sons and daughters are infertile. There are no reliable statistics. Many women who have had several miscarriages had been exposed to DES. The drug also triggered malformed reproductive tracts, such as a cervix that does not shut tightly to hold the growing baby—an incompetent cervix. To be sure, thousands of DES sons and daughters, the lucky ones, were spared the toxic side effects. Yet virtually

every woman exposed to DES has vaginal adenosis, benign cellular changes that can be precancerous and that require constant
monitoring.

It goes without saying that physical defects belie the harsh psychological wounds for mothers, daughters, and sons. Mothers who
thought they did everything they could to maintain a healthy pregnancy were creating girls who would never have the same experience. The tragedy, wrote Drs. Roberta Apfel and Susan Fisher in *To
Do No Harm*, is that the very people who are supposed to protect
you—doctors and mothers—were the ones who caused harm. "It
is the providers of early tender care who have let down the DES
daughters and sons—unwittingly, the mothers brought pain to
their children; the physicians brought pain to their patients." As
Pat Cody, a DES mother and health activist, once told a reporter,
"You can't imagine what it's like to think that with the best of
intentions, you might have poisoned your daughter; her life might
be in danger because of something you did. You blame yourself.
You can't imagine facing your daughter and telling her."

What is even more troubling about the DES saga is that even
when solid evidence proved that the drug was not effective, it continued to be given for years. In 1959, DES was banned for use in
poultry, but not for use in humans.* The drug was finally pulled in
1971 after a few oncologists proved the cancer connection. The
other dangers would emerge years later.

What were women thinking then? Why would anyone pop a

---

* DES was given to chickens and cattle to make them grow tastier meat faster.
For a while it was a boon for the chicken industry, like speeding up the factory line—
that is, until a few poultry-plant workers who munched on DES-laced chicken necks
got big breasts and became sterile. Years later, science writer Nicholas Wade would
call DES a "chemical of bizarre and far-reaching properties, chief of which is that
it is a spectacularly dangerous carcinogen." (science, 1972) But in 1959, few people
were thinking about the DES-exposed human babies.

pill while pregnant without thinking about the long-term effects on the baby? Today, every pregnant woman scrutinizes labels and surfs the Web before drinking diet soda. What was going on?

One of the going childbirth myths from antiquity right up to the 1950s was that dangerous things did not pass through the placenta or breast milk. That was nature's gift to the perpetuation of the human race, or so it was thought. "There was always the assumption that the placenta was protective. People called it the placental barrier," said Dr. Roy Pitkin, UCLA professor emeritus of obstetrics and gynecology. The term itself implies protection. Your fetus was *barricaded* within the placenta, safe from all the nasty things in the outside world. That helps explain why so many women continued to smoke during pregnancy. *Woman's Home Companion* told readers that alcohol and cigarettes "play no significant role" in pregnancy.

> "People called it 'the placental barrier.'"

There had been scientific studies demonstrating just the opposite—that tiny molecules, the kind that make up most medicines, slip easily through the placenta. But that did not change the way doctors practiced or how women thought about their pregnant bodies. Even in the nineteenth century, a handful of savvy doctors noticed newborns with traces of lead or mercury, common treatments that had been given to their pregnant mothers. Francois Magendie, a French physiologist, injected camphor into pregnant dogs and saw the drug in the fetuses. W. Reitz, a German doctor, gave mercuric sulfite to pregnant rabbits and found the drug in the fetal brains. Working in reverse, W. S. Savory, an English scientist, injected strychnine into fetal dogs, and the pregnant mothers convulsed. C. C. Huter, a professor of obstetrics at the University of Marburg, found chloroform—used as an anesthetic—in the umbilical cord blood of newborns. "Huter's studies should have alerted physicians but they did not," remarked Donald Caton, an obstetric anesthesiologist and author of *What a Blessing She Had Chloroform: The Medical and Social Response to the Pain of Childbirth from 1800 to the Present*. A quarter of a century later, Paul Zweifel, a Swiss physician, conducted scrupulous experiments to prove that when a mother uses chloroform or even takes salicylic acid (aspirin), it gets into the fetal blood system.

Besides the fact that we all thought the placenta was like a Berlin wall around our babies, we had faith in science then. DES peaked in popularity in the 1950s when Americans were the most optimistic we have ever been and ever will be. We considered ourselves victors in war and victors in the laboratory. We were heading to the moon, combating Communism, and ridding the world of dangerous germs. We became a consumer nation and believed that man-made topped nature. Formula was better than breast milk. McDonald's was better than home cooking. Our doctors were Marcus Welbys, the most trusted men on the planet. When a doctor told you he had a pill that would prevent a miscarriage, he believed he was helping you and you believed him. We were not naive. We had faith in the system.

DES emerged from the same basic science and the same hormone enthusiasm that went into making the oral contraceptive. Drs. George van Siclen Smith and Olive Watkins Smith, a brilliant Harvard husband-and-wife team who promoted DES, worked in the same unit at Harvard as John Rock (who did the first human trials of the pill) and across town from Gregory Pincus (who did many of the first animal studies). In 1960, when the FDA approved the pill but restricted its use to a few years, fearing long-term dangerous side effects, Ortho Research Foundation's medical director turned to the Smiths, the DES folks, when he wanted an expert letter sent to the FDA to urge them to remove the restrictions.

In fact, in 1946, when investigators were figuring out what to do with the synthetic hormones, Dr. Fuller Albright, a Harvard endocrinologist, suggested that DES be used as an oral contraceptive, or, as he called it, "birth control by hormonal therapy." His far-fetched idea—which would later be referred to as "Albright's prophecy"—was tucked into a few paragraphs on page 966 of an internal medicine textbook. Given all the excitement in the Harvard halls about DES and how it ever so gently tilted hormone levels, Albright proposed daily DES pills to prevent ovulation and induce menstruation on the "least undesirable day." Ultimately, a pill was created but with a different concoction of hormones that juggled our hormones in a slightly different way.

The gung-ho atmosphere for using drugs to improve pregnancy or to prevent it ushered in another drug fiasco, thalidomide, which is often confused with DES. Thalidomide was used to prevent morning sickness but was discovered in the early 1960s to cause physical deformities in babies, such as missing or flipper-like arms and legs. Thanks to Dr. Frances Kelsey, a cautious FDA officer, it was never approved in the United States, but it was used overseas—and by a few American wives who were living abroad. The thalidomide saga shocked the world because it proved, for the first time, that drugs crossed the placenta. It also proved that animal testing does not always predict what will happen in humans. Thalidomide is a safe drug for pregnant rats.

Still, DES babies looked healthy. The thought of dangers emerging 15 or 20 years down the road seemed absurd. And while a few doctors warned about the cancer-causing potential, most women read glowing stories. The way hormones were talked about, they did not seem like medicine at all. It was natural, restoring the body to its original balance. It was in this atmosphere that the pill was approved, and by the 1960s, 6.5 million American women were taking it. And it was in this atmosphere that DES was born and nurtured.

In 1938, British scientist Sir Charles Dodds trimmed two of estrogen's unique four rings and concocted the first synthetic estrogen. His imitation estrogen was more potent than the real stuff, a trait that frightened Dodds but thrilled the rest of the world. "We still have to proceed with caution on any long-term hormonal treatment of the human female," Dodds warned. Years later, he would tell the late health activist Barbara Seaman that "we should always be humbled when we think of what we do not know about the female reproductive cycle."

"We should always be humbled when we think of what we do not know about the female reproductive cycle."

Nevertheless, Dodds published his recipe for mock estrogen in the widely circulated journal *Nature* for all to see, knowing all too well that his American colleagues would be eager to mass-produce it. It's been said that Dodds rushed to publication to trump German scientists and quash their ability to capture the lion's share of the market and use the profits to feed the Nazi regime. Whatever the impetus, by 1938 the instruction manual to make synthetic estrogen was out there for all to see. Within a few years, there were at least 200 different brand names for DES vying for a share of the market. The best part for drug makers was that synthetic estrogen was mass-produced easily. Real estrogen, by contrast, was collected from mare's urine, which was labor intensive and expensive. The best part for women was that DES was an oral medication. Real estrogen was by injection only.

Several doctors refused to offer it to patients. The Council on Pharmacy and Chemistry—a cautious committee of the American Medical Association—said that "because the product is so potent and because of the possibility of harm, the Council is of the opinion that it should not be recognized for general use or for inclusion in New and Nonofficial remedies at the present time and that its use by the general medical profession should not be undertaken until further studies have led to a better understanding of the proper functioning of such drugs." *Good Housekeeping* warned readers that stilbestrol "is new and highly controversial. Conservative obstetricians feel that their value is as yet far from proven."

The official AMA statement pointed to the possibility that DES could cause cancer. "The possibility of carcinoma induced by estrogens cannot be ignored. . . . Other tissues of the body may react in an undesirable manner when the doses are excessive and over too long a period. This point should be firmly established, since it appears likely that in the future the medical profession may be importuned to prescribe to patients large doses of high potency estrogens, such as stilbestrol, because of the ease of administration of these preparations."

The *Journal of the American Medical Association*, the AMA's mouthpiece, published an article by three prestigious researchers who reviewed the dangers of synthetic estrogen that emerged in animal studies: rats with liver necrosis; mice with sudden paralysis and death; rodents with hemorrhagic spleens. "It would be safer to confine the drug to experimental use for the present until the significance of the side effects is better understood," concluded Drs. Ephraim Shorr, Georgios N. Papanicolaou, and Frank Robinson.*

Despite the few doomsayers, the underlying theories supporting DES made sense for everyone else. The Smiths at Harvard believed that women miscarried because their bodies did not produce enough progesterone in the first weeks of pregnancy, an idea that still holds sway among infertility specialists. DES increased progesterone because estrogen triggers the body to make more

* Papanicolaou had devised the Pap test for cervical cancer in the 1940s.

progesterone. "Stilbestrol is not given because it is estrogenic but because it stimulates the secretion of estrogen and progesterone," Olive Smith once said to correct her critics, who claimed that she was giving the drug for a supposed estrogen deficit.

In 1941, the FDA gave the green light for limited use. DES was approved to treat vaginitis and gonorrhea; to reduce menopausal symptoms; and to suppress lactation. As soon as DES hit the shelves, doctors prescribed it for all sorts of things outside of the four approved uses. The FDA advises, but doctors can do whatever they please with available drugs. Pharmaceutical sales reps, the point people who go door to door, urged doctors to use DES for irregular periods, painful periods, morning sickness, and high-risk pregnancies, to name a few off-label uses.

In 1947, based primarily on the Smith studies, the FDA okayed DES for use during pregnancy. Some women who took DES had already had a miscarriage and were desperate to do anything to maintain their next pregnancy. Some women never had a miscarriage but figured an ounce of prevention was worth a pound of cure. Some women didn't even know they were taking DES. It was added to their prenatal vitamins unbeknownst to them. DES was also given to women to prevent dangerously high blood pressure in pregnancy, and it was given to diabetic women, who are at high risk for pregnancy complications. Many women were getting a lifetime's worth of estrogen packed into nine months, essentially consuming the equivalent of a nine-month stock of birth control pills every day. One DES mother figured out that she swallowed more than 10,000 milligrams of the stuff, "roughly the equivalent of 500,000 of the current low-dose birth control pills."

One Smith study suggested that DES babies were bigger and healthier than babies born to women who did not get the drug. The results were picked up immediately by marketers. "Yes, desPlex to prevent abortion, miscarriage and premature labor,"

"Really?"

Yes...

des*PLEX*®

to prevent ABORTION, MISCARRIAGE and PREMATURE LABOR

recommended for routine prophylaxis in ALL pregnancies . . .

Grant Chemical Company advertisement for desPlex (DES pills) in the June 1957 issue of the American Journal of Obstetrics and Gynecology. Courtesy of DES Action.

cheered a popular magazine advertisement that showed a chubby allegedly DES baby.

According to Dr. Frederic Frigoletto, former chief of obstetrics and gynecology at Massachusetts General Hospital in Boston, odds are that after a woman miscarries, she is likely to carry the next baby to term regardless of treatment. Give her a pill and she will thank the pill for her good fortune. He never prescribed DES but remembers the initial enthusiasm. "The culture was different," he said. "People thought the doctor was always right and the doctor thought he was always right. There was much less in the way of patient awareness. There was much less rigidity in the scientific process and so it was a lot of opinion and intellectual decision making. The DES trial probably would never have passed muster in today's world," when hospital review boards must approve every trial for safety, efficacy, and proper informed consent.

Dr. William J. Dieckmann was a hardheaded, skeptical midwesterner who was not easily swayed by savvy marketers or prestigious scientists. He was a Harvard-trained physician and knew the Smiths well. He went on to become chief of University of Chicago's Lying-In Hospital. In the early 1950s, he conducted a study with the intention of confirming the Smiths' findings. He felt that a study with a control group—comparing DES takers to other women—was necessary.

The Smiths were delighted with the prospect of a large study that they suspected would confirm their work and convince the naysayers. Several doctors had already written to the Harvard team saying they could not duplicate their positive DES findings, to which Olive Smith would always reply that they most likely had not followed the "Smith and Smith protocol to the T." The Smiths had a detailed system that included starting the drug very early in pregnancy with low doses of DES and building up as the pregnancy progressed. If DES pills were started a day too late, it would not work. If a woman forgot to take a pill one day, it destroyed the delicate hormonal juggling. The Smiths offered the Chicago team advice and told Dieckmann that they wished they had had a control group of women who got placebos. "We are, of course, all agog, to know what comes of your study. We're so glad that your controls got placebos, that was the big flaw in our series."

Dieckmann presented his results at the American Gynecologi-

> "Our rationale for stilbestrol is something more than a philosophy."

cal Society meeting in Lake Placid in 1953. He had compared 860 women taking DES with 806 women taking placebo. To minimize bias, neither doctor nor patient knew who got the dummy pills until the end of the study. Everyone received the same kind of prenatal care. He said that his study showed that DES was ineffective. He published his findings in the November issue of the *American Journal of Obstetrics and Gynecology*. Scientific journals tend to be written in a dry, just-the-facts style. After the chummy correspondence during the study, Dieckmann's paper included a biting critique of the Smith's original study. The journal published an equally angry rebuttal by the Smiths. This was about as close as you'll get to a shouting match among the typically tight-lipped doctors.

Dieckmann said that no other researchers could reproduce the Smiths findings. Was he hinting that their data was fudged or sloppy? And there was more. He trashed their methodology, particularly because the Smiths did not have a control group. "Preventing miscarriage by the daily consumption of a few tablets is indeed enticing, however a careful perusal of the literature reveals that most of the clinical data are not supported by adequate controls." He snidely included step-by-step instructions of how a properly controlled trial should be done. The Dieckmann study showed slightly higher rates of miscarriages among women getting the real thing, but he admitted that the numbers were too small to blame the drug. He believed that his research proved that DES was not effective. He did not say DES was dangerous.

In response, Olive Smith quipped, "Our rationale for stilbestrol [Eli Lilly's brand of DES] is something more than a philosophy." She and her husband attacked the Dieckmann trial for comparing a heterogeneous group of women. They pointed out that the Harvard study was limited to high-risk women who benefit from the drug. Dieckmann included a variety of patients. They added a pompous kicker claiming that the messy University of Chicago study would never have gotten the okay from Harvard statisticians. "Our experience with the use of stilbestrol continues to be satisfactory and to confirm our previously reported clinical trials," they wrote. "We never claimed it was a panacea, but after ten years of careful study and observations of patients with bad and even hopeless prognoses at the onset of pregnancy when stilbestrol was started we are convinced that the many obstetricians who have been following our recommendations for the use of stilbestrol in pregnancy will realize that the paper presented this morning . . . fails to provide definite evidence to the contrary."

Harvard's prestige trumped the University of Chicago's scrupulousness. Drug companies were eager to help Harvard promote their positive findings. The Smiths were not only believers; they were confident and persuasive. George Smith was a physician who had treated hundreds of women who seemed to deliver healthy DES babies. Olive Smith was a talented biochemist who thought she understood precisely why the drug worked. She once told a doctor who worried about the ramifications of synthetic hormones that she realized that the ever-so-slight modifications from the real thing could spark dramatic changes in the body, for better or for worse. "The synthetic hormones are a tricky lot. We are just plain lucky with stilbestrol in pregnancy, I think." Because the Chicago study found DES ineffective but not dangerous, some doctors believed they were better off giving the pill until further proof emerged.

There were concerned scientists. In the 1960s, after DES had been used for years, an Australian doctor wrote a letter to the Smiths saying that his experiments with ewe showed that chronic and high doses of estrogen triggered horrible diseases. And while animals are not always good predictors of human behavior, he conceded, he did wonder whether the Smiths had any long-term follow-up of their patients. He wondered about the possibility that DES could prompt cancers years after it was given. "It seems presumptuous of me to make this suggestion," he wrote, "but you will recall that it is only recently that the relationship between lung cancer and smoking in urban areas has been established. I suppose I might add that some 500 years ago, all the experts were convinced that the American continent did not exist; and even Christopher Columbus got a shock!" Olive Smith replied that while they did not trace their patients, she was sure the drug was safe. Her arguments were so convincing that the Australian doctor wrote back asking for samples to start treating his patients.

For women, the chances of getting DES depended more on where her doctor trained than what the studies found. Doctors who trained under the Smiths were more likely to prescribe DES than doctors elsewhere. UCLA's Dr. Roy Pitkin remembers his mentors at the University of Iowa talking about Dieckmann's "absolutely marvelous study. It was not randomized, but it was blinded [neither doctors nor patients knew whether they got the drug or dummy pill]. The answer was very clear. [DES] did not do any better than doing nothing at all."

Dr. Pitkin added that the Chicago study was not as persuasive as the investigators hoped. "People must have said, 'I know that research showed it won't do any good, but it can't cause harm.' People felt they had to do something and the people in Boston say it really works. Patients at that time were going to do what their doctors advised and the usage [of DES] had a clear geographical distribution. If you were from Boston where it was promulgated, it continued to be used. On the other hand, I was at the University of Iowa and I was taught that this was a drug thought to be useful but found out not to be so we did not think it should be used. So the adverse effects were geographical too."

Dieckmann would not come out unscathed either. Years later

he would be sued for misconduct in his trial because participants claimed they were never told they were being given a hormone. They thought they were testing a prenatal vitamin.

Unless you have had a miscarriage, you cannot imagine the profound sadness that hits a woman who loses a baby. One day you are pregnant, feeling the world is smiling at you. You plan your life as a mother and imagine the child-to-be. The next day, often times after a massive hemorrhage, the reality of motherhood vanishes. Your dream becomes a nightmare. You worry that you will never be able to maintain a pregnancy.

"Once upon a time it was believed that a woman who had had three miscarriages stood almost no chance of producing a live baby. But now the picture is different."

If you had miscarried in the 1940s or 1950s and you read articles about a drug that promised big healthy babies, you, too, may have been lured by its promises. "Once upon a time it was believed that a woman who had had three miscarriages stood almost no chance of producing a live baby. But now the picture is different," said *Woman's Home Companion*. "Doctors have tests to detect a shortage of hormones. Today, from 74 to 90 percent of women who have had repeated miscarriages can get through this dangerous period with hormone treatment."

Despite studies suggesting that estrogen spurred cancer and that DES may not be effective, the babies looked fine. Everyone, it seemed, had friends who took the drug and saw their seemingly healthy babies. Some women went to doctors who refused to give them DES, so they switched doctors. The rhetoric was grandiose. Thirty-nine out of 42 women who had suffered from several miscarriages had successful pregnancies after taking daily doses of estrogen, *Time* magazine reported. "Scientific progress is being made in discovering ways to avoid miscarriages . . . *put yourself in the hands of your obstetrician early,"* a woman's magazine extolled. "I know there is no guarantee against miscarriages," another magazine writer concluded, "but I'd like to feel that I'm doing everything possible to prevent it."

In 1968, Cynthia Laitman Orenberg, who had lost her first pregnancy to miscarriage, was a medical researcher and wary of taking an estrogen-stimulating pill during the next pregnancy: "Although I didn't know a great deal about hormones, I knew enough to understand they had wide-ranging effects on the body." She also knew about recent medical fiascoes that showed that women had to be wary of drugs during gestation. The headlines had just hit the papers that cigarette smoking was dangerous to the fetus. And she knew about the recent thalidomide fiasco. Orenberg's doctor reassured her that DES was perfectly safe. "Pregnant women have been taking them for at least 20 years, and believe me, nothing happens to them or their babies," she recalled him saying. She was convinced.

Between 1966 and 1969, a rare form of vaginal cancer called adenocarcinoma of the vagina struck seven young women. When vaginal cancer does occur, it typically strikes the elderly. In this cluster, the oldest patient was 22, the youngest was 15. All of them were treated at Massachusetts General Hospital.

"There can no longer be doubt that synthetic estrogens are absolutely contraindicated in pregnancy."

Doctors were baffled. Some of the girls had originally been misdiagnosed because doctors never considered cancer of the vagina striking such young women. Some had just started menstruating, and when they complained of excessive bleeding, the doctors assumed they had had irregular periods, typical among young teens. When Joyce Bichler, a DES daughter and vaginal cancer survivor, went to her health clinic, they examined her and accused her of having a botched abortion.

At first glance, the girls with cancer seemed to have nothing whatsoever in common—that is, until one mother asked her doctor whether DES, a drug she took during pregnancy, could have anything to do with her daughter's cancer. Never before had a drug taken during pregnancy been tied to defects years down the road. Her concerns could have easily been dismissed as the crazy talk of an anxious mother. This woman had no medical background, no scientific rationale for such a claim, and she knew nothing about pharmacology. But Dr. Arthur Herbst, a gynecologic oncologist, took her seriously. Rightly so.

Herbst, who had trained at Harvard and treated these women, said he had a hunch the mother may be on the right track. He had read a 1964 study that showed that mice whose mothers were given DES were at increased risk of cancer. The author of that little-publicized study concluded, "We feel that abnormal hormonal environments during early postnatal (and antenatal) life should not be underestimated as to their possible contribution to abnormal changes of neoplastic significance in later life."

In some ways, it would seem preposterous that a drug taken during pregnancy could promote cancers and deformities years later. But basic embryology lends credence to these notions. Every embryo is like a mini-repeat of human evolution on fast-forward. The newly created human transforms from speck to person within nine months. As early as the first five weeks or so, a series of ducts snaking through the lizard-like embryo twist, expand, and clamp shut in a precisely timed ballet as they remake themselves into new body parts. What seemed like a simple tube will become the birth canal, or the ovaries, or the lungs. The choreography depends, in part, on the delicate balance of hormones bathing the baby-to-be. At specific intervals during the formation of the baby's reproductive tract, the fetus is exquisitely sensitive to the balance of male and female hormones. A slight tweak in the system can trigger grave repercussions. No wonder DES given early in pregnancy wreaked the most havoc, more so than DES given later. The estrogen boost fiddled with the natural balance. (Some scientists and DES daughters wonder whether the early impact of DES on the fetus has altered genes that could transmit dangers from generation to generation. So far the claim has not been backed by conclusive studies, nor has it held up in a court of law.)

Dr. Herbst, now the Joseph B. DeLee Distinguished Professor emeritus at the University of Chicago, and his colleagues conducted a thorough analysis of the seven cancer victims. They showed for the first time a link between women who took DES and cancer in their daughters. His findings were published on April 22, 1971, in the *New England Journal of Medicine*. Shortly thereafter, Dr. Peter Greenwald confirmed the findings. Greenwald examined five New York teenagers with vaginal cancer. The cancer killed three of the women. The other two survived after radical surgery. One woman had her vagina removed, the other a vaginectomy plus hysterectomy. Just as in Herbst's study, all the mothers took DES during pregnancy. No mothers in a control group of daughters without cancer had taken DES. "There can no longer be doubt that synthetic estrogens are absolutely contraindicated in pregnancy," Greenwald concluded. "This raises concern about the future. It is not known how many more vaginal carcinomas will be developing, whether other types of cancer will also develop in men or women, or whether a longer induction period may be present for those who received a smaller dose."

Greenwald, who is currently director of the Division of Cancer Prevention at the National Cancer Institute, said that immediately after his study, he wrote a letter to the *New York Times* alerting physicians about the dangers of DES. The point was not only to alert doctors to stop giving the drug, but to warn women who had taken it to have their daughters examined for vaginal cancer. Caught early, vaginal cancer is curable. Neglected, it kills. Greenwald was called to testify during congressional hearings about DES. "The FDA seemed not to be convinced at first," he told me, "saying there is just one study. I said there are two studies with solid verifications."

After days of the contentious hearing, the FDA reversed its DES decision. In 1971, months after the studies were published, the FDA banned DES during pregnancy. Though sales plummeted, not everyone believed the hullabaloo over a drug that had been used for so long. Many doctors and women were worried that a good drug was being taken off the market. Further studies would shatter that idea.

"DES daughters are the only ones who don't need to send their mothers a Mother's Day card."

DES would be blamed for even more harmful consequences. The warnings, not surprisingly, did not end the story of DES, but launched a new episode. DES sparked patient activism, scientific studies, and a slew of lawsuits.

Friday morning, April 23, 1971. Pat Cody poured herself a cup of coffee and opened her *San Francisco Chronicle*:

DRUG PASSES CANCER TO DAUGHTER

A synthetic hormone that was given to seven women during pregnancy has apparently caused a rare vaginal cancer in their daughters up to 22 years later, Harvard medical scientists reported yesterday.

The scientists theorize that the drug—stilbestrol—somehow, they don't know how, altered the development of the vagina, allowing the cancerous growth.

"I thought, oh my God, I took that. But I tried to rationalize the fears saying the cancer was rare," Cody said. She wasn't alone. It was hard to miss the news that day. Andrea Goldstein was a senior in high school in a suburb outside of Boston. She remembered that morning because there was a hole in the *Boston Globe* where her mother had ripped out a front-page article. Goldstein, who has since become a passionate DES activist, went to the library to find out what her mother was hiding from her. "Rare Cancer Linked to Synthetic Hormone." She learned that she, too, was a DES daughter.

The Herbst study was reported in every major newspaper. Thousands of women, like Cody, were probably doing the same thing that morning, having their morning caffeine and reading headlines such as

FDA WARNING ON SYNTHETIC ESTROGEN

GIRLS CANCER LAID TO MOTHER'S DRUG

RARE CANCER TYPE LINKED TO A DRUG

And like Cody, they probably hadn't given their prenatal care a moment's thought in the past few decades.

Susan Helmrich, an epidemiologist and a DES daughter, believes that DES triggered her carcinoid lung tumor, though proof is lacking. Helmrich has lived through her own DES hell but considers herself one of the lucky ones because "I've managed to live my life." She was diagnosed with vaginal cancer at 21. They constructed a new vagina with her colon, which worked well, but the intestinal surgery triggered several bouts of bowel obstruction. "It was awful, horrible. I've had seven surgeries, lung cancer, and three breast biopsies. I know you have to weigh the risks and benefits with every drug, but with DES, the data were so

clear. The Dieckmann study that came out two years before I was
born was clear as day. It showed DES was completely ineffective,
yet it was marketed for another 18 years."

Pat Cody was, without a doubt, an unfortunate victim of a
spectacular medical fiasco. And yet in a bizarre way, Americans
are lucky that of all the women affected by DES, Cody, a rabble-
rousing feminist activist, was one of them. (She and her husband,
Fred, opened the famed left-wing Cody's bookstore in Berkeley.)
Cody was not going to let the DES story die. She just switched
gears, or veered ever so slightly, from left-wing rebel to health
activist.

What began as a few friends meeting around her kitchen table
blossomed into a global DES activist organization that still exists
today.* "If you sent out the letters in the 1950s, women would
have thrown them out," she said, but not so in the 1970s. Cody
joined forces with women like Andrea Goldstein, who were rally-
ing women on the East Coast. They fought for legislative changes
that increased funding for DES research, published newsletters
encouraging women to have their daughters examined for DES-
related reproductive changes, and pushed TV executives to incorporate DES into story lines. (The *Mary Tyler Moore Show* had a DES
episode). Cody bombarded *Saturday Night Live* with hate mail when
Jane Curtin, during a 1971 eve-of-Mother's Day Weekend Update,
said, "DES daughters are the only ones who don't need to send
their mothers a Mother's Day card."

There have been perhaps hundreds of lawsuits; but just as
many have never made it to court. It was not an easy process.†

---

* For more DES information, contact DES Action USA:
Web: www.desaction.org
e-mail: desaction@comcast.net
toll free: 800-337-9288
† Among other hurdles, many women were stymied by statute of limitations
law. Sybil Shainwald, a lawyer who specializes in women's health issues, was a

In the early years, some cases were thrown out of court because the suits were filed too many years after the alleged injury. A few states reformed the law so that women had three years after the time of the discovery of the injury, not after the use of the drug. Also, with more than 200 brands of DES, few women could remember which product they took. All too often, women took a mix of brands, depending on what their pharmacist stocked that day. On top of that, most women could not get access to their 20-year-old patient records. And even if they did, sometimes the DES information was missing. For some doctors, it was a routine drug, like prenatal vitamins, so it wasn't included in the medical records. Although DES has been tied to a variety of reproductive ailments, it is difficult to prove that the pill itself is responsible for a woman's infertility or her miscarriages. Needless to say, lawsuits may be cathartic for some women, harrowing for others.

In the early 1980s, a few lawyers fought for legal changes (passed in several states, including California, New York, Wisconsin, and Michigan) that allowed women to sue every drug company that made DES if they could not identify their specific pill. If the plaintiff won, the drug companies would be responsible according to the market share they had during the woman's pregnancy. The early wins made headlines and once again, DES was in the news.

Without a doubt, DES is a story that comes and goes from the

---

leading litigator who brought about many changes that helped women win cases against DES makers. New York, for instance, has a general limitation of three years. According to Judge Jack B. Weinstein, Senior U.S. district judge, who wrote an opinion on the *Braune v. Abbott* case in 1995, "As part of a general 'tort reform' package [in 1986], the New York State legislature adopted a 'discovery' rule for torts involving substances with latent effects. . . . Under the new statute, the three-year limitation period runs from the time that that plaintiff discovered, or would have discovered, her 'injury.'" J. B. Weinstein, memorandum and order in *Braune v. Abbott Laboratories, et. al.*, Eastern District of New York, United States District Court, 1995.

media's eye—first in the 1950s as the wonder drug, then in the 1970s as a cancer-causing agent, and then in the 1980s as the focus of massive malpractice suits. "Those of us who were affected by it were Flavor-of-the-Month about 20 years ago," wrote columnist Anna Quindlen, a DES daughter, in a 1993 *New York Times* Mother's Day piece. Some say the drug itself changed the nature of the patient-gynecologist relationship, crystallizing and politicizing anger brewing among American women.

Part Four

10

From Kitchen-Table Surgery to the Art of the C-Section

In the summer of 1979, Dr. Florence Haseltine, an obstetrician, had a cesarean section because she wanted to. She planned it that way even though there was no medical reason for the surgery. No one else on staff thought it was a good idea. But Haseltine decided "she wanted to have a baby, not a delivery." She had seen plenty of women in labor to know what she was in for had she gone the vaginal route; and she had done enough C-sections to know what the operation entailed.

When Haseltine was pregnant in the 1970s, women were Lamaze-breathing their babies out while husbands coached on the sidelines. In the you-go-girl atmosphere of *Our Bodies Ourselves* and Title IX, it was the odd woman who asked for more technology and who wanted more men around. (About 90 percent of obstetricians then were men, so chances are if you asked for anesthesia or surgery you were asking for more men in the labor room.) Even odder that Haseltine was a die-hard feminist. (Among her long list of accolades, she would go on to found the Society for Women's Health Research, sit on the board of American Women in Science, and earn the American Woman's Medical Association Scientist Award.)

Haseltine does not consider drug-free, technology-free births an extension of women's rights. Quite the contrary. Feminism, for her, means acquiring sufficient information to understand the medical choices and demand the one that suits you. A few of her colleagues remember quite clearly when Haseltine asked for her C-section. They said that her decision was unusual for the time but not for Haseltine, a savvy woman never swayed by popular opinion.

"You get pregnant to have a child, not to have a delivery," Haseltine said. "I could not understand everyone's angst over my C-section. The baby was cooked. Get her out. My husband thought it was fine. He was furious with the people who were against it." She said she felt safer with a surgical delivery, avoiding the complications of vaginal births, such as tears and hemorrhage. Thinking about the tiny yet swelling group of women nowadays who request C-sections for no reason other than that's what they want, you have to wonder whether Haseltine was ahead of her time or a harbinger of bad things to come.

Today the cesarean section is the most common operation in America among women of reproductive age. Hysterectomies come in second. In 2005, nearly one in three babies were born surgically, a record high. When Haseltine gave birth in 1979, about 15 percent of babies born in the United States came out via C-section, and doctors called that an epidemic. (Doctors also cried "epidemic" in the 1930s when the C-section rate climbed to 2 percent.)

Most women do not have Haseltine's medical know-how, nor have they had the opportunity to watch several vaginal deliveries and several C-sections to know exactly what they are getting themselves into. Most women are not demanding C-sections just because they want them. The vast majority of women having C-sections are having surgery because their doctors told them the procedure is medically best.

For most of us—even if we surf the medical Web pages and read pregnancy advice books—we can't help but be swayed by the way our doctors talk to us. When our doctor encourages us to try to give birth vaginally and offers reassuring statistics, we feel confident that we can do it. When a doctor lists everything that could go wrong in natural childbirth and regales the hospital's triumphant surgical record, we feel scared to push. Let's face it, we are all aiming for a healthy baby. Naturally, we want the simplest and safest choice, but what one woman considers simple is another woman's toil. The crux of the C-section debate is figuring out how much of our doctor's advice stems from a concern about our health and that of our baby and how much stems from their fear of litigation or a desire to fit the birth into a tight schedule. What is knowable about the harms and benefits of the available paths? And how much risk is too much? How's a woman to know? The few women asking for C-sections for no medical reason adds another layer to the delivery debates. Who's in charge here?

Julius Caesar was not born by cesarean section. (And neither, by the way, did he have anything to do with Caesar salads.)* Scholars doubt the C-section lore because women who had C-sections in ancient times died afterward. They either bled to death or succumbed to infection. Caesar's mom lived to see her son rule, which makes Caesar's cesarean unlikely.

Why the Caesar moniker? There are two theories. According to one, the operation was named after an ancient Roman law, *Lex Regia*, which later became known as *Lex Cesaria*—as in Caesar was

---

* As for the salad, that history is even more contentious. According to legend, it was invented in 1924 by Caesar Cardini, an Italian chef in Tijuana.

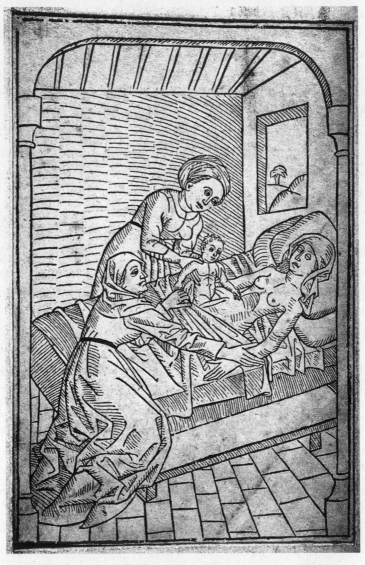

Woodcut of a baby being delivered by cesarean section of a dying mother. It is one of the first obstetric scenes in a printed book. From the woodcut book Seelwurzgarten, *printed by Conrad Dinkmut, Ulm, 1483.* Courtesy of the Wellcome Library, London.

the ruler, not as in Caesar was born that way. The law, enacted in 715 BC (long before Caesar's birth by whatever means) mandated that the fetus be removed from a dead pregnant woman prior to burial. You can't help but wonder why Caesar would have insisted on pinning his name on a burial procedure; surely there were more prestigious laws he could have picked. That's why the other theory is more believable, which states that *cesarean* comes from the Latin *caesuru*, "to cut."

Before the late 1800s, cesarean sections were death rituals, not lifesaving procedures. If a doctor suggested a C-section, you knew you were on the way to the morgue. To save time, sometimes priests operated so they could take out the fetus and baptize it immediately. Try picturing a C-section of the Middle Ages: Women were wide awake, tethered to the table and cut without anesthesia. Sometimes they were not sewn back together because doctors thought an open wound hastened healing. According to midwives, cesarean sections were necessary when the patient "does not respond to penetrating odors, is ice-cold, without pulse, looks collapsed and pale as death, and if her breath leaves no traces on a mirror." In other words, they were ready for surgery when they looked dead.

The first recorded successful C-section was in 1508. Jacob Nufer, a Swiss pig gelder, operated on his wife. Supposedly, after 13 midwives tried to get the baby out to no avail, he grabbed one of his knives and got his baby. His wife survived and gave birth five more times. Many scholars have a hunch that the pig gelder tale is about as believable as Caesar's mother's C-section. They do not believe that Nufer's wife could have survived the amateur

If she "does not respond to penetrating odors, is ice-cold, without pulse, looks collapsed and pale as death, and if her breath leaves no traces on a mirror."

operation and then survive five more vaginal deliveries (including a set of twins) without rupturing her uterus. They also wonder why no one said anything about Nufer's operation until 82 years after the event. You would think the farmer would have been so proud of his surgery he would have spread the word.

In what is perhaps the most clever marketing pitch ever, the name was changed sometime in the sixteenth century from "cesarean operation" to "cesarean section." That took the edge off, as "section" sounds so much simpler than "operation."

Even more mind-boggling than letting your farmer-husband slice you (if you believe the story) are the few reports of women who did their own C-sections. Most of the time the baby died. In 1822, a 14-year-old Long Island maid pregnant with twins buried herself in a pile of snow and cut open her belly and pulled out one baby. Her boss, a doctor, found her and delivered the other baby and sewed her back together. The teen mom survived, but there is no record of the babies. In 1879, a Turkish woman who had been in labor for 36 hours used a razor to get her baby out. Her neighbor stitched her. Mom and baby survived. It's been said that the do-it-yourself operations were safer than having a doctor do it because doctors carried germs from other patients.

No one knows who did the first professional C-section in America. Dr. John Lambert Richmond, a janitor and Baptist minister who attended enough medical lectures to earn an MD degree, did one on April 22, 1827, using "common pocket instruments." The woman developed an oozing infection, but she and her baby survived. Dr. Richmond is honored in Cincinnati, Ohio, with a commemorative statue that states: "In Memory of John Lambert Richmond, MD, who performed the first Caesarean Section in America." Maybe so.

But maybe not. According to another tale, Dr. Jesse Bennett delivered his own child by C-section on January 14, 1794. Supposedly, he laid his laboring wife on planks balanced across two barrels and dosed her with "a large dose of laudanum" (an opium derivative), took out the baby, and took out his wife's ovaries, too, so she would not get pregnant again. Some people doubt the Bennett story. They think that his wife had the C-section but it was done by Alexander Humphreys, a respected obstetrician.

By the early years of the 1900s, a quartet of events converged to transform cesarean sections from a deadly encounter to a viable and increasingly popular option. The discovery of germs encouraged doctors to wash their hands and prevent potentially life-threatening infections.* The gradual acceptance of anesthetics in obstetrics enhanced the appeal. The rise of hospital births gave doctors more leverage. Lastly, the need for intervention skyrocketed. Rampant malnutrition striking the recent influx of immigrants produced bone deformities and caused many women to have misshapen pelvises that blocked the baby's exit.

> "The women of the large urban centers have become steadily more insistent in their demands for shorter and less painful parturition."

In *Lying-In*, Richard and Dorothy Wertz credit German physician Max Sanger for boosting survival rates and popularizing the operation. In 1882, he used aseptic technique and silk thread and

---

* The germ theory has been called the "bacterial revolution," as if the discovery was like a coup, causing a major upheaval in the way doctors treated patients. Nothing works that way in medicine. It's more like evolution than revolution. Scientists discover something in the laboratory, and after a dangerous lag time, doctors adopt the new notions in the hospital. Michael Worboys, in his book *Spreading Germs*, argues convincingly that there was a long and dangerous lag time between the laboratory discoveries and clinical practice. Michael Worboys, *Spreading Germs: Disease Theories and Practice in Britain, 1865–1900* (New York: Cambridge University Press, 2000).

claimed that 80 percent of his patients survived. It was an astonishing triumph compared with previous survival rates that were somewhere between 30 and 50 percent.* Todd Savitt, in *Medicine and Slavery*, credits, or rather blames, Southern doctors for perfecting the operation because they practiced and demonstrated on "volunteer" slaves much the way they perfected the technique for sewing torn vaginas by operating hundreds of times on postpartum slaves. In 1828, for instance, Mrs. Payne, an African-American Virginian, had her C-section in front of several doctors and "leading citizens" who watched her and her baby die shortly thereafter.

In 1908, Franklin S. Newell, a Harvard physician, put forth the prescient notion of C-sections for social reasons, maintaining that elite turn-of-the-century ladies should have C-sections because they were too fragile to push. He published his comments in an article called "The Effect of Overcivilization on Maternity." He said packed social calendars left women no time for fresh air and exercise, which weakened their strength to push. Poor people were tougher because they played outside. He must have forgotten about sweatshops. The solution to the modern woman's ails, Newell facetiously suggested, was to revamp society or to change obstetrics to cater to the physical demise of women. He opted for the latter:

> The advocacy of an elective Cesarean section for patients
> without pelvic obstruction will undoubtedly come as a shock
> to many members of the profession who have heretofore

* A 20 percent death rate from C-sections seems grim by today's standards, but it was a huge leap for those days. Just a few decades earlier, an 1858 British study of 80 C-sections, yielded a 29 percent survival rate. Among 120 operations in America, between 1852 and 1880, 42 percent survived.

considered obstetrics as a science to be conducted by rule of thumb according to traditions laid down by past generations. Let any man, however, who has had an extensive practice among the women of the overcivilized class honestly consider patients he has seen, whose health has been wrecked by so-called conservatism, and who subsequently required a more or less serious operation to restore them to even a moderate degree of health, and ask himself the question whether these patients would not have been better off if some method had been employed at the time of delivery which though seemingly rash, would have prevented not only the nervous exhaustion of prolonged labor, but the damage to the pelvic organs which resulted from one of the so-called conservative operations.

Newell's paper seemed silly then. More women were having surgical births, but doctors did not encourage them and they were certainly not suggesting that healthy women without signs of pelvic obstruction undergo risky surgery. As rates creeped upward, though, doctors began to worry. Critics accused doctors of doing surgery for their convenience (to avoid being up all night with a woman in labor) or to make money (because you could charge more for operations than for vaginal births). They also accused women of asking for risky surgery to avoid labor pain. "The women of the large urban centers have become steadily more insistent in their demands for shorter and less painful parturition, and the accoucheur may disregard these demands only at great risk to his own practice, " said New York City public health officials.

A 1933 New York City maternal mortality investigation blamed C-sections, in part, for maternal and fetal deaths. They worried about the inordinately high C-section rate. It was 2.2 percent of all births. They said that doctors opted for surgery because "the performance of a caesarean section is technically less demanding than the more difficult vaginal delivery. Consequently, it is subjected to misuse and pressed into service when better judgment and greater skill would permit delivery by the less hazardous normal route." They pointed to the New York Lying-In, where the C-section rate was 4.98 percent among private patients compared with 3.28 percent among the rest. (The same is true today, where numbers are almost ten times higher among private patients.)

Before antibiotics, doctors developed techniques to keep the womb away from the rest of the body to prevent the spread of germs. The Waters technique, developed by Dr. Edward Gilmay Waters, pushed the bladder down and moved everything that lay above the uterus so the womb could have its own breathing space. After antibiotics, the issues were no longer about where to put the womb to keep it away from the rest of the body, but how to cut the woman. Did she need the long vertical incision, which created a large opening, making it easier for the doctor to grab the baby; or could she get away with a low transverse cut, what doctor's call the Pfannenstiel incision (after its creator) but women call the bikini cut.

Leslie, who works in marketing for an environmental think tank, was 33 years old when she had her first C-section because, as she put it, the baby was breech and sunny-side up (he was bum-first, not headfirst). The doctors tried to turn him, but he wouldn't budge. He was too big. He weighed more than 10 pounds.

"Once a Caesarean always a Caesarean."

Leslie wanted to go vaginal for the next births. But by the summer of 2007, vaginal births after cesarean sections were increasingly rare. For years, doctors had stuck to the words of advice from a famous turn-of-the-century obstetrician, Joseph DeLee, who once decreed, "Once a Caesarean always a Caesarean." His word was dogma. The fear has always been that once a uterus is cut, that slice—no matter how much or how little scar tissue—could burst during subsequent vaginal deliveries. In the 1980s, when women were fighting against increasing rates of C-sections, many women started having vaginal births after C-sections with no problems at all. The process became so commonplace that it was dubbed VBAC (pronounced "veeback"), for vaginal birth after cesarean. During the heyday of VBAC, the national C-section rate dropped ever so slightly, reflecting the number of women not having second or third C-sections. The trend has slowed in recent years, as studies have shown that the risks of vaginal births after surgical ones are rare but real.

Leslie wanted to experience birth and avoid surgery, but her doctor, like so many doctors nowadays, convinced her otherwise. "I was doing more than reading books at Barnes and Noble. I was reading the *New England Journal of Medicine* articles about VBAC. In 2004, in an article Leslie most likely read, the *New England Journal of Medicine* published the results of a 33,000-woman study that found that the rate of uterine rupture from VBACs was 0.7 percent, fewer than one in 100 cases. For some women, that's a risk not worth taking. For others, the numbers are so low that they'd want to give vaginal birth a shot, hoping that should things go awry, their doctor can switch to surgery at the last minute. The study found that about three-quarters, 73 percent, of women who wanted to deliver vaginally after C-sections had successful deliveries. The rest had to have last-minute C-sections. Emergency C-sections are riskier than those planned in advance. Investigators also found that women who tried to have VBACs were more likely to need blood transfusions and were more likely to have endometriosis (where bits of the uterine lining fall outside the uterus), compared to women who had repeat C-sections.

Despite the potentially scary statistics, Leslie said she thought a "vaginal birth would be better for me, for my whole body. That is the natural way to have a child. Your body kicks into gear. Your baby comes out of the birth canal, not the uterus with a cut." But Leslie's doctor quashed any hopes of a vaginal delivery when he told her that she could deliver vaginally but "you're basically signing your life away." (He may not have used those exact words, she added, but that was the sentiment expressed.)

The American College of Obstetricians and Gynecologists does not encourage VBACs. The organization recommends that if a woman insists on trying to deliver vaginally after a previous surgical birth, the hospital should have a surgical team "immediately available" during the birth; that means paying an obstetric surgeon, a nurse, and an anesthesiologist to stand by just in case. Some hospitals may not deem the understudies worth it. From a cost analysis, you may as well have the surgeons and anesthesiologists do something if they are going to stand there. As for Leslie, she said that all things considered, she would have preferred a vaginal birth, but she is not going to harp on the loss of one. "It is what it is and unfortunately, this is my reality. I don't feel like, 'Oh my God, I'm yearning for that [vaginal delivery].' I would have felt badly if I never breast-fed my children."

> "By the time they found out that people are a little more people-friendly than machines, it was too late."

Fetal monitors record the baby's heartbeat and display it on a screen in the labor and delivery room and in the nurse's station. It is routine. The concern among many childbirth advocates is that any blip on the fetal monitoring screen—anything that could seem slightly worrisome, even if it has no real outcome data attached—prompts doctors to surgery. Their fear is that should anything go wrong with the vaginal delivery, the monitor strip will end up in a court of law where the judge will not be able to distinguish the meaningless blip from an ominous one. Studies suggest that when fetal monitors first came into use, C-sections skyrocketed. Maureen Corry, executive director of Childbirth Connection, said it is "mind-boggling" that women today would have to hunt for the rare provider to agree not to do fetal monitoring. Her concern is that the monitor kicks in a cascade of interventions that are likely to lead to a C-section. Once a woman is attached to the machine, she is confined to a bed and on her back, which is thought to slow labor. The slow labor encourages doctors to use drugs to induce labor. Sometimes, drugs—if used too soon in high doses—can cause strong contractions too early, another reason for surgery.

As Dr. Frederick Naftolin, New York University's head of reproductive biology research, put it, "We took this tool, electronic fetal monitoring, that had a lot of advantages and applied it wholesale. It was never meant to be applied wholesale. Then hospitals and administrators said we don't need as many people, these machines can do it. By the time they found out that people are a little more people-friendly than machines, it was too late. The nurses were already transferred to other wards."

The point of the fetal monitor was to detect babies who are under stress—say, lacking oxygen—and get them out as soon as possible. A large meta-analysis of nine randomized controlled trials, including about 53,000 women, found that continuous fetal monitoring was no better than using a stethoscope to hear the heart rate every so often. Naftolin said that several studies have shown that if a woman has a soothing caregiver by her side, labors tend to be shorter. "There is reason to believe," he added, "to let a woman enjoy as much freedom [of movement] as she can."

"Too Posh to Push."

When Victoria Posh planned her C-section for no medical reason (just the way Haseltine had), British tabloids screamed,

"Too Posh to Push." It was a snappy headline used ad nauseam
to describe allegedly soaring rates of cesarean section by maternal
request. In reality, few women want surgery for no medical reason.
Doctors say the rate, which is based on murky statistics, is about
1 percent. According to the survey *Listening to Mothers*, only one
woman among 252 respondents requested a C-section without
medical reason.

Still, the media attention frightened doctors. Both the American College of Obstetricians and Gynecologists and the National
Institutes of Health have each had consensus panels to decide
what to do with the woman who insists on surgery. Can a woman
tell a doctor to give her an operation? The answer is not clear-cut.
Dr. Charles Lockwood, chairman of Yale University's Department of Obstetrics, Gynecology and Reproductive Sciences, says,
"If an individual believes the bulk of the literature that an elective
C-section is as safe as a trial of labor then he or she is obligated
to respect the patient's autonomy and perform a C-section. However, if their interpretation of the literature indicates that a trial of
labor is safer, than their obligation is to first do not harm, which
will lead them to refuse the request." His remarks echo recommendations of the American College of Obstetricians and Gynecologists, which put the difficult decision back in the hands of the
doctor and patient. In other words, no one is saying whether this
is a good thing or a bad thing. The truth is that women who can
afford to choose their doctor will opt for one who caters to their
wishes. If you want a natural childbirth, go to a doctor who will
give you one. And if you want a C-section, it just takes a quick
Google search to figure out if your doctor has high rates of surgery. As some surgical adocates see it, if a woman can opt for
cosmetic surgery, why shouldn't she be allowed to ask for surgery
to deliver a baby?

No matter how you slice the data, it is hard to believe that one in three women are unfit for vaginal deliveries. Cesarean section rates have increased by 46 percent in the past ten years without a comparable drop in maternal mortality. Are we saving lives? And if not, are we sparing some women dangerous consequences of birth? Ideally, scientists should figure out which woman is likely to tear from a vaginal delivery or which baby is likely to get stuck and then do surgery on only that small targeted group. No matter how much technology we add to the delivery room, it does not seem to take out the guesswork. It only leads to more operations.

"How you view childbirth is a reflection of your philosophy of life."

Dr. Naftolin thinks that hospitals should revive medical forums where doctors discuss every C-section. Perhaps a discussion among peers, rather than with lawyers, would shift the decision tree. At Mt. Sinai Hospital Medical Center in Chicago, researchers initiated a two-year program where every woman had to get a second opinion before surgery, criteria were established for C-sections, and every case was discussed. The results, published in the *New England Journal of Medicine* in 1988, found that C-section rates dropped from 17.5 percent of 1,697 deliveries in 1985 to 11.5 percent of 2,301 deliveries in 1987.

The tricky part about the overall statistics and the committees and the guidelines is that they ignore the uniqueness of each woman going through labor. There are so many things to consider when patients and doctors grapple with the decision, including the size and health of the baby and how many children the woman plans to have (each subsequent C-section raises the risk of complications).

And yet, the C-section debates are not simply calculating the odds of things going wrong. They tap into your basic concept of childbirth. Are you focused on the journey as much as the destination? If you are worried about surgery, you must consider strongly whether your physician truly believes that a C-section is the best route or whether he is trying to avoid potential litigation.

Time and again, childbirth studies teach us that there are so many secrets yet to be learned, so many things that prevent us from streamlining the process of giving birth. Why does labor occur on this day and not that one? What happens to our chemistry, the chemicals that seep into the baby when our body convulses in labor? There is speculation—but just speculation—that the adrenaline-like hormones that surge during labor are good for the baby. Maybe they add the necessary behind-the-scenes final touches before the baby enters the world's stage. As Marcie K. Richardson, an obstetrician at Harvard Vanguard Medical Associates and an instructor at Harvard Medical School, said, "How you view childbirth is a reflection of your philosophy of life. One of the things I feel really sad about is our culture where the message is that women can't have a good childbirth experience without turning themselves over to the control and interventions of the medical community. Starting parenthood with the notion that somebody else has to manage the process is not empowering. This is something that I think about, but I suspect many of my colleagues would say, 'What is she talking about?' I was really proud of my own ability to birth my 9-pound 13-ounce son vaginally without anesthesia. I remember touching his head at my vaginal opening and holding him on the delivery table and being in love. I was his mother. I am not opposed to C-sections, but I'm pretty sure a 30 to 40 percent rate is not justified."

11

Freebirthers

John Shanley was born—or as his mother, Laura, says, "flew out"—on August 21, 1978, sometime around 1:30 in the morning. Laura was on all fours on her bed in Boulder, Colorado. His father, David, caught him. Rick, a friend, cut the umbilical cord. Ron, a filmmaker, captured it on tape. Bill, Laura's ex-boyfriend, was also there, along with his new boyfriend, Bruce. There was no doctor. No midwife. In fact, no one in the house had any formal training in childbirth at all. John came into the world precisely as his parents planned—together with family and friends and yet, from a medical standpoint, all alone.

Laura would give birth four more times the same way—no doctor and no midwife. (Except that the following births would be minus Rick, Bill, Bruce, and Ron). Laura, David, and others of their ilk—there are hundreds of like-minded couples around the world—call their way of delivery unassisted childbirth, and they call themselves freebirthers, as in free of all medical intervention. They say it's healthier physically and psychologically for the baby to enter the outside world in a soothing, loving place without the glare of hospital lights, noisy nurses, intrusive doctors, and meddlesome midwives poking about.

Do-it-yourself childbirthers are either one end of the spectrum or a group of outliers. No one can say for sure. The numbers, whatever they are, are small. Even if there are hundreds of women who have given birth this way and hundreds more who are planning on it, that is nothing compared with the 365,000 babies born every day. Numbers aside, one thing is certain. Do-it-yourself deliveries demonstrate, albeit in an extreme way, that many women are angry about high-tech pregnancies. The movement (if you can call it that) reflects a group of women who are concerned about the perceived callous care and the potential impact on babies. Maureen Corry, director of Childbirth Connections, a not-for-profit organization that aims to improve the quality of maternity care, says that unassisted births represent an "indictment of the whole maternity care system. When women's choices are not honored and respected, women will go to extremes to have the kind of birth they want."*

Freebirthers say they are not having babies this way to prove they can do it. Most of them believe they are doing what's best for the child because a newborn's first moments of life are ingrained in the psyche. If you enter the world with smiles and cuddles, you'll be a happier grown-up than if your first glimpse of the world outside the womb is latex gloves and feet in stirrups.†

---

* Childbirth Connections was formally Maternity Care Association, a 90-year-old birth education organization, one of America's first groups to launch the natural childbirth movement, to invite Grantly Dick-Read to the United States to speak, and to train midwives and promote their care. They do not condone unassisted births.

† A few psychologists believe that emotional influences begin before birth, so if people are saying nasty things to you or you are spending your entire pregnancy yelling at people, your child will come out angry and stressed. The good news is

Other freebirthers are not concerned with the psychology of it all but opt for do-it-yourself childbirth because they see no reason for doctors or midwives in the first place. Matthew Jasper, from Framington, New Hampshire, had never heard anything about do-it-yourself deliver- "Hey there's ies, but after the relatively easy births of his lots of red in first two children, he said to his wife, "Next there!" time why don't we do this by ourselves." And so they did. Athena Burke, another freebirther, moved from Boston to rural Petersburgh, New York, to give birth to her first child, Christopher, in a tent in her backyard so he could "be born among the big hug of the mountains and listening to the birds and the water flowing as his first sounds."* Natalie Picone-Louro said she "opted out of prenatal care because I trusted my body. I didn't want the whole peeing in a cup, doing the heart rate, it all seemed so unnecessary."† Prenatal exams "took away from the beauty of

---

that you can go to a rebirther, who can generate your before-birth memories and help you heal. The other good news is that if you believe in all this mind-body stuff and worry about how it will affect the offspring, and are planning on becoming pregnant and are afraid to do a do-it-yourself birth, you can think happy thoughts during pregnancy and then go to the hospital for the delivery, though freebirthers may not condone that. It's just deciding how much you are considering the psychological fetal impact versus psychological moment-of-delivery impact versus no impact at all.

* In Judy Seaman's documentary *A Clear Road to Birth*, Athena's elderly neighbor Victoria Green said the town was flabbergasted. "People couldn't believe it. If it had been the other way around, they could have believed it better—that a *Petersburgh* girl could have moved to Boston and had her baby in the backyard. But for a *Boston* girl to come to Petersburgh and have her baby in the backyard was astounding."

† You pee in a cup because doctors are worried about gestational diabetes, which strikes about 5 percent of all pregnant women. As in ordinary diabetes, the body's system of converting glucose from food to energy has gone awry. High levels of glucose have dangerous complications for the mother and baby. Babies tend to be born very big, making deliveries difficult. Mothers are at risk for preeclampsia

being pregnant and I wanted to be in control." Her toddler Trinity watched. "I had a birth tub in the kitchen [an inflatable toddler pool] and the whole house was lit with candles. It was really amazing. At one point, I lifted my pelvis. . . . I reached through my legs and brought her into the water . . . and Trinity said, 'baby you did it. You did it baby!'"

No one knows for sure how many women freebirth—the sobriquet sounds more extreme sport than childbirth technique. There is a veritable feast of Web sites, including labouroflove.com and empoweredchildbirth.com, to name a few. Laura's Web site is called Bornfree (www.unassistedbirth.com). She is proud to say that if you Google childbirth—not even unassisted childbirth, but just plain childbirth—her site ranks sixth. (I did it, and it came out eighth—still extraordinarily popular for an unpopular way to have a baby. But two years later it dropped to 33rd). They share freebirthing stories and advice, such as how to cut an umbilical cord (dental tape and sterilized scissors, though you can also wait until the cord falls off the placenta by itself, allowing the baby to get a few hours or days of extra nutrients).

They also share videos. One shows the big sister and dad climbing into the baby pool right after mom delivered the baby in it. Everyone is giggling. "Hey there's lots of red in there!" shouted the sister, right before she took off her clothes and climbed in. Seeing the videos and talking to women who have gone the unas-

---

(high blood pressure and protein in the urine). According to the American College of Obstetricians and Gynecologists, gestational diabetes is more likely to occur in women who are obese, sedentary, and have high blood pressure, high cholesterol, and a family history of diabetes or who have polycystic ovary syndrome. Many freebirthers insist that they stay healthy, exercise, and eat right during pregnancy, minimizing the risk of diabetes and other birth complications. But according to ACOG, while unfit and overweight women are at increased risk, gestational diabetes can strike any pregnant woman.

sisted route is so inspiring. They all have such upbeat stories of their birth. It's such a warm, cozy affair. Toddlers are watching. Sometimes friends are hanging out in the kitchen. Moms never shout nasty things to their husbands, the way some of my friends told me they did during their hospital deliveries. In the videos of freebirthing, laboring women are cooing, not hollering. They talk in breathy voices and say things such as, "Come on baby, we love you." Sometimes the mom and newborn are in bed. Sometimes they are in a bath. Spectators are gleeful. Everybody is encouraging. Babies slip out. Moms are calm. Blissful is a better word. Dads are helpful.

Freebirthers have had two conferences. About 40 couples showed up for the first one in 1998 in Charleston, South Carolina. Some 50 adults along with their babies attended the second one in July 2001 in Louisville, Colorado. Rixa Freeze, for her PhD thesis, interviewed more than 200 women who delivered their own babies successfully.

Still, statistics are unreliable because the data do not differentiate between an *intentionally* unassisted birth from an *unintentionally* unassisted birth, the kind born in a taxi or airplane. Some freebirthers worry they may be breaking a law, so after the baby is born, they get a licensed midwife to sign the birth certificate. In some states, such as New York, it is illegal to give birth with an unlicensed midwife but not by yourself. Do-it-yourself deliveries are not illegal precisely because it would be impossible to prove that a woman intentionally chose that path.

Safety records are misleading because some women intend to deliver unassisted but if things are not going as smoothly as planned will rush to a hospital and call a midwife. Judy Seaman, the writer, producer, and director of the documentary *A Clear Road to Birth*, said that of 54 women who planned unassisted births, 2 called a midwife and 8 went to the hospital at the last minute. The

rest had successful births. The records, if there were any, would
have shown 44 successful unassisted births.

The neighbors "think I'm not only a boob but a dangerous one at that."

Long before the Web sites, videos, and freebirthing associations, there was one lone mother who gave birth to seven of her nine children without any help at all. In the 1950s, Pat Carter, of Titusville, Florida, refused to have doctors and midwives help her. Birth is easy, she said, "if you don't let yourself gain too much weight and the whiskey helps you relax." According to the wire service story that ran August 20, 1956, "Mrs. Ellerbe Carter sipped several highballs, then delivered her seventh child unassisted at 1:20 this morning. By 9 a.m. she was back at work in her husband's real estate office with her 6-pound son at her side."

Carter invited a reporter to watch, hoping it would help the League of Liberated Women, a floundering antidoctor crusade she started. In what may have been the only thing her one-woman organization did, she promised to send a diploma to anyone who said they gave birth without medical assistance. It said:

> Because she chose to, and did, bear a child without the attendance of physician, nurse, or midwife, which evidences liberation from the false belief now generally prevalent in our Time and Society, namely that the presence of a professional birth attendant is necessary or desirable even in cases where no mental or emotional disease is present in the Mother, and where no physical disease or deformity exists in the mother or child, and where no malpresentation or malproportion exists. And BECAUSE she has thus escaped from the rules, regulations, restrictions, and humiliations

Mrs. Carter Has Done It Again

Delivers Ninth Child — Herself

By MARY LOU CULBERTSON
From The News-Journal Bureau

TITUSVILLE—Mrs. Ellerbe W. Carter has done it again.

Mrs. Carter gave birth to her ninth child at 1:20 a. m. this morning unassisted by doctor or midwife. She made no outcry and she handled the proceedings with neatness and dispatch.

The six pound boy, named William Douglass, was her third son. Mrs. Carter gave birth to the child in the bedroom of her home. Her two next to youngest children, Teddy, 3, and Claire, 14 months, slept peacefully in another bed in the room during the birth.

Several hours prior to the birth Mrs. Carter sipped whisky highballs. She appeared to be cold sober, however, as she tied the cord and chatted gaily about the wonders and joys of bringing a baby into the World "as nature intended."

She explained that the whisky supplied the relaxation needed for a natural birth.

After the cord was tied, the baby dressed and sleeping cozily in its crib, Mrs. Carter announced its arrival to the rest of the family

Pat Carter makes headlines in the Daytona Beach News-Journal *for delivering her baby without any medical help.* Reprinted with permission by the The Daytona Beach *News-Journal.*

now generally endured by women in childbirth when 'conducted' in accordance with prevailing birth mares [I assume she meant mores].

There are no records of anyone ever being awarded a certificate. It goes without saying that in 1956 BC (Before Cyberspace), spreading a message was tough. Pat's birth story made the local papers, and she even got into *Look* magazine, where she was quoted as saying that the neighbors "think I'm not only a boob but a dangerous one at that." Ironically, the same issue spotlighted actress and new mother Janet Leigh on the cover, who proclaimed that she was "so groggy, I don't remember much about this."

CERTIFICATE of EMANCIPATION
and
Lifetime Membership Award

BECAUSE she chose to, and did, bear a child without the attendance of physician, nurse, or midwife, which evidences liberation from the false belief now generally prevalent in our Time and Society, namely that the presence of a professional birth attendant is necessary or desirable even in cases where no mental or emotional disease is present in the Mother, and where no physical disease or deformity exists in mother or child, and where no malpresentation or malproportion exists;

And BECAUSE she has thus escaped from the rules, regulations, restrictions and humiliations now generally endured by women in childbirth when "conducted" in accordance with prevailing birth mares:

IT IS OUR CONSIDERED OPINION THAT. . . . . is therefore entitled to THIS CERTIFICATE OF EMANCIPATION.

And BECAUSE of her emancipation as aforesaid, we hereby award her LIFETIME MEMBERSHIP in THE LEAGUE OF LIBERATED WOMEN'

IN TESTIMONY WHEREOF, witness the will of this League through the signature of its duly authorized officer this _____ day of _____ , 19____ .

LEAGUE OF LIBERATED WOMEN
By _Katherine Machennes_
President

Pat Carter's certificate honoring other women who gave birth without any medical help. Courtesy of Pat Carter and her daughter, Mary Winn.

In her self-published 370-page manifesto, *Come Gently, Sweet Lucinda,* Carter said she decided to call her theory of birth *euthagenesis* (Greek *eu* for "good" or "well" and *genesis,* for "origin.") The word *eugenics* was already taken, so she added a "tha," which has no meaning at all.* Or as Carter would put it, eugenics had come to mean "something a little snobbish, like picking out grandsires and grandams with IQs of 300 or thereabouts." Not surprisingly, her way of birth with the tongue-twister name did not create a sensation or even make a wave. It was barely pronounceable.

* As Yale historian Daniel Kevles wrote in *In the Name of Eugenics,* the name was coined by Francis Galton in 1883 to mean "good in birth." But by the "first half of the twentieth century, eugenic aims merged with misinterpretations of the new science of genetics to help produce cruelly oppressive and, in the era of the Nazis, barbarous social results."

Euthagenics, as she pointed out, never made it into Webster's dictionary. It's remains, today, one of the few un-Google-able terms. (If you try to Google euthagenesis, Google responds, "Do you mean euthanasia?")

Carter sprinkled her book with snippets of peculiar and dangerous advice. She told women to starve during pregnancy to "prevent the pooch," otherwise called a pregnant belly. She told women to smoke, drink highballs, and minimize calcium so the baby's bones would be soft and slide out of the birth canal with ease.* She recommended boned corsets, or as she put it, "BONED, B-O-N-E-D. . . . This will really stop the little rascal."

Like the freebirthers today, she gave do-it-yourself advice. About cutting the umbilical cord, she wrote, "Anyone who can tie their own shoe strings can do it." She told women not to wipe off the amniotic sac because, she claimed, it prevented diaper rash. As for the afterbirth, she wrote, "It is soft and floppy, and if you press it, it will flop over into a fold. And zip. There it is."†

---

* That may sound bizarre by today's standards but many women were put on strict diets then, not only for the supposed health of the baby, but to stay skinny. "The girl who's neither underweight nor overweight can have an even more glamorous figure *after* pregnancy than before," according to a Harvard doctor writing about a pregnancy diet in *McCall's*. ("If You're Eating For Two," F. Stare, March 1956, *McCall's*)

† She said *placenta* comes from the Latin word for "cake," and she assumed "they made pancakes in those days." Some freebirthers today bury the placenta. The placenta has its own weird history and alleged attributes. In *History of Childbirth*, French historian Jacques Gelis says that one ancient ritual included rubbing the newborn's face with its own placenta to guarantee a smooth complexion. And along the lines of Carter's insinuations, there was a time when the placenta was thought to confer special fertile powers and has been used from time to time as an ingredient to help infertile couples. For a few hundred years beginning in the sixteenth century, according to Gelis, European explorers wrote home about various tribes who practiced so-called placentophage, eating ripe placentas. J. Gelis, *History of Childbirth: Fertility, Pregnancy, and Birth in Early Modern Europe* (Boston: Northeastern University Press, 1991).

Nowadays, freebirthers like to say that they are giving birth the way nature intended, the way most animals do. If they had their druthers, cats prefer to give birth alone. So do many other animals, but not all. Many animals, like most humans, like companionship. Mice like to have female mice around. Female dolphins swim around their parturient dolphin friends and nudge the newborn to the surface for its first breath of fresh air while the new mother rests underwater. Elephants huddle around the laboring elephant mom, swaying from side to side with her. They're either comforting her or keeping dangerous predators away. When contractions start, one elephant friend waits by the birth canal and helps the baby out—all 250 pounds of it. Then another one flings the amniotic sac into the air and lets it float to the ground, where it becomes a mat for the baby. Father birds help feed the newborns, but orangutan dads are the best helpers of all. When the newborn is crowning, the dads stick their big fleshy lips to the newborn's head and suction the little guy out of the birth canal like a vacuum extractor.

"Primate Labor when observed does not seem excessively difficult."

All of these births, from a human gaze, happen so easily and so comfortably. "Primate labor when observed does not seem excessively difficult," writes Harvard University anthropologist Peter Ellison in his delightfully entertaining book *On Fertile Ground*. "The female may pace and change positions frequently, but then usually delivers in a squatting position without assistance, often delivering the placenta herself by pulling the cord or allowing it to fall out later as she moves about."

So why the ruckus when a few *Homo sapiens* want to go back to nature and give birth the way our fellow mammals do: undisturbed, relaxed, and, more often than not, alone? The fundamental philosophy of freebirthers is that female humans would give birth more easily if, like their nonhuman primate friends, they chilled out and if, again like their nonhuman primate friends, they were not surrounded by all the fuss of medical monitoring and doctors and midwives. Except for the part about being alone, the underlying premise is remarkably similar to that of Lamaze founder Elisabeth Bing and her disciples, who encouraged women to calm down. The difference—and it is a colossal one—is that the folks who push natural childbirth are encouraging women to give birth *without* drugs but *with* assistance.

Dr. Sarah Buckley is an Australian doctor, freebirther, and author of *Gentle Birth, Gentle Mothering: The Wisdom and Science of Gentle Choices in Pregnancy, Birth and Parenting*. She believes that when all species of females are in labor, they are hyperalert for danger, an evolutionary survival tactic to prevent enemies from eating the new baby. This basic instinct, the same drive that fought off predators, triggers an adrenaline rush when a laboring woman enters the hospital and gets a whiff of a strange odor or hears a sudden noise.

There is some tantalizing evidence to suggest she may be right with regard to the stress-and-labor connection. A Guatemalan study compared women who labored alone with women who had doulas (companions by their side during labor). The upshot: women with doulas had shorter labors, 8.8 hours versus 19.3 hours, and were less likely to have cesarean sections, 19 percent versus 27 percent. The assumption is that the doulas helped the women relax, which in turn shortened their labor and prevented complications. Animal observations suggest similar conclusions. Apes and monkeys, it seems, may be able to switch on or off contractions on a whim. Such animal tales have been reported by scientists trying to watch a monkey give birth, only to discover that the contractions stopped while the monkey was being observed, and then the baby was born as soon as investigators left the room—

even if the observers left for just a few minutes. Was it a conscious decision by the parturient mother, or did the nosey scientist trigger an involuntary rise in the expectant mother's stress hormones? The point, according to freebirthers, is that stress screws up the birthing process, which may be why so many hospitalized parturient women inevitably need drugs to jump-start labor and speed it along. The drugs are overriding our natural instincts.

But there is another side to the argument that has more to do with the physical than the emotional. It has everything to do with the simple fact that we are humans. Try as we might to calm down during contractions, laboring humans are not going to give birth as easily as apes and monkeys. Our hips are narrower and our babies' heads are bigger.

When we stood up and got smarter a few million years ago, birth got tougher. Our pelvises got narrower so we could walk upright rather than dangle like apes, and our heads got bigger to hold our thoughts. That made for a tight squeeze at birth. If our hips were—heaven forbid—wide enough to allow a baby to glide out, we'd be waddling and tripping. According to a compelling theory put forth by Wenda Trevathan, a professor of anthropology at New Mexico State University, the narrow-pelvis-big-brain shape may have prompted some of our great-great-great-great-grandmothers to ask for help—a sort of survival of the fittest kind of thing. It's not just the cramped exit, said Trevathan. She points to the fact that due to the shape of the pelvis, ape babies slide out in a straight path and face their mothers, so the ape mom can reach in between her legs, see if her newborn is breathing, and carry it up to her breasts. In contrast, due to a different sort of human-shaped pelvis, our babies come out facing away from us, making it more of a gymnastic feat to reach the baby, lift it out, and worry about the breathing and a potential umbilical cord around the neck.

Trevathan has a hunch that our larger-brained ancestors (compared with their ape grandmothers) were smart enough to realize that their babies had a better chance of surviving if they had a helping hand. As she puts it, "It's hard for me to imagine that someone said, 'my goodness, my baby is coming out backwards, I need help.' But I think we became conscious of our vulnerability as our brains became larger and we had a certain amount of fear— I hate to use that word because I don't think it's close to terror. Some uncertainty and anxiety that led them to seek companionship. We are gregarious creatures and by going to another person for assistance when we did begin to deliver the baby, we knew we had someone there if there were some kind of complications. It's not as Draconian as if you don't get help, you won't deliver. If you look back in history, even a tiny difference in life or death would make a big difference, the numbers would begin to pay off in time, to become statistically significant. Probably several millions of years ago, the benefits of having a person around outweighed the negatives."

Unfortunately, Trevathan explains, what began as a helper who probably respected the birthing woman's input into the whole thing has evolved in many cases into an attendant who really does not care what the woman giving birth wants. In a chapter she wrote in Robbie Davis-Floyd's *Childbirth and Authoritative Knowledge*, Trevathan concludes that "for millions of years the birthing female was the most important member of the 'obstetrical team,' but today her knowledge about her body is often suppressed and 'managed.'" It was a slippery slope of one hominid asking another hominid for assistance to a twenty-first-century pregnant woman letting her doctor do whatever she or he felt necessary.

On the Web, many freebirthers write about the head rush when the baby hits their G-spot, the surge of hormones as the baby knocks into vaginal nerves. In an article in *Mothering* magazine

called "They Don't Call It a Peak Experience for Nothing," Ruth
Claire wrote that she was expecting the "dogging pain" of labor
but was shocked by the "sensation of sexual ecstasy, the voluptuous feeling of penetration. . . . Crouched
on my knees on a little afghan, I caught the
infant who rushed from my vagina into the
small world between my legs, in the midst
of an extraordinary orgasm from the inside
out." I am envious. Debra Pascali-Bonaro,
a midwife, made a documentary called *Orgasmic Birth*. Women,
shockingly so, allowed her to film their births, which were supposedly orgasmic, as the name implies. (Some of the women did
not look like they were having orgasms, but looked the way you
would expect a woman to look giving birth naturally.)

> "They don't
> call it a peak
> experience for
> nothing."

Freebirthers insist that if making a baby is a sexual experience,
delivering it should be, too. It's a motto that makes sense at first,
but on further consideration seems a dubious argument. Making
dinner and eating dinner are two very different culinary experiences. Also, when you are making a baby, you are in bed with
your lover ostensibly becoming aroused and potentially creating
a family (unless you are in a fertility clinic having sperm shoved
into you by someone you just met). Nine months later, your uterus
is contracting and your vagina is stretching to inconceivable pro-
portions. Most likely, you are surrounded by a crowd of gawkers,
otherwise known as birth attendants. Sometimes people whom
you have never met are shoving fingers in your vagina. Getting
pregnant can be romantic. Getting the baby out is anything but.

But that is precisely why freebirthers say naysayers have it all
wrong. If women gave birth at home in bed with a lover by their
side, they, too, could experience the highs of childbirth. Think of
it the other way around, they insist: Try making a baby in a hospital
bed with physicians and medical students watching and comment-

ing. Or as Laura Shanley put it, "Imagine the difference between having sex in a hospital surrounded by hospital personnel and machinery and having sex at home in a candle-lit bedroom. Take away the drugs and machinery, take away the watchful concerned eyes, take away the fear, and a whole new world opens up to us."

Laura Shanley did not have an orgasm during delivery, but she had one a few minutes before when she had sex during labor—another astonishing accomplishment performed by a freebirther.

Shanley insists that freebirthers are not a "bunch of earthy-crunchy granola types," though the home videos of unassisted birthers would suggest otherwise. (Many freebirthers homeschool their children and refuse to vaccinate them.) They definitely appear more Birkenstock than Prada. Shanley insists that she is not an "anti-doctor fanatic." She gets angry when articles (there have been many) claim that she and others in the unassisted birth community believe

The American College of Obstetricians and Gynecologists is against all homebirths, even with a midwife.

that birth is inherently safe provided it is not interfered with by doctors and midwives. She says, "That to me is only one piece of the puzzle. There is poverty and inside interference—fear, shame and guilt."

Their logic is not far-fetched. Sometimes doctors or midwives do cause problems. And stress has been shown to tighten muscles and increase pain. But then again, their logic is not completely accurate either. Sometimes doctors save lives. Either way, freebirthers believe that if you get rid of intrusive birth attendants, stay nourished, and banish fear, you've scored a hat trick when it comes to birthing. In *The Power of Pleasurable Birth*, freebirther Laurie A. Morgan writes that unassisted birth "can be one of the safest, most responsible choices out there." Chances are you and

your baby will be fine, because birth complications are not the
norm. But you are taking a risk. Shanley's fourth baby was born
five weeks premature and died after delivery. She believes that the
baby, who had a congenital heart defect, would have died despite
hospital care. And perhaps, she said, the baby was better off dying
peacefully at home rather than hooked up to lifesaving machinery
at a hospital that may have extended its life for only a few weeks.

Freebirthers say—though virtually every doctor would
disagree—that neonatal and maternal mortality rates are abysmal
in developing countries *not* because of a lack of top-notch doctors
and access to hospitals, but because the women are malnourished.
Harvard's Peter Ellison, who has written extensively about childbirth, sees it otherwise: "In the developing world, and particularly
in rural areas where people practice subsistence agriculture and
horticulture with little interaction with the larger national market
economy and its services, virtually all mortality has a nutritional
component. But the sources of maternal mortality that are the
most important are nearly always puerperal fever and hemorrhage.
Does malnutrition make mortality from puerperal infection more
likely? Yes. But would antibiotics and more sterile conditions help?
Absolutely."

Needless to say, midwives and doctors are not condoning birth
without assistance. The American College of Obstetricians and
Gynecologists is against all home births, even with a midwife.
About 5 percent of parturient women require blood transfusions,
which means they were hemorrhaging, says Dr. Sarah Kilpatrick,
who chaired the American College of Obstetricians and Gynecologists committee on obstetric practice and is head of the Department of Obstetrics at the University of Illinois. In Great Britain,
the Royal College of Obstetricians and Gynecologists, responding
to a spate of news stories, issued a statement about unassisted
births. They said that women should have the right to give birth

in a comfortable environment, but "the practice of freebirth is new to the UK and little research exists regarding its safety and success."

Freebirthers, naturally, would reply that stress and doctors caused the problems in the first place and prenatal visits are useless. It's tricky to say how many visits are warranted. Indeed, prenatal care is a relatively modern birthing phenomenon, one that came with the birth of obstetrics as a profession. The American College of Obstetricians and Gynecologists recommends monthly visits for the first 28 weeks, then every other week for the next 8 weeks, and then once a week until the baby is born. That's about 14 visits on average for the low-risk woman. As Dr. Joshua Copel, a professor of obstetrics and gynecology at Yale University, says, "What we hope to accomplish is a long talk [about prenatal care]. Exactly how we improve outcomes nobody really knows." A 2001 World Health Organization study that reviewed data from more than 55,000 women found that fewer prenatal visits compared with standard treatment did not make any difference at all when it came to maternal mortality, urinary tract infections, preeclampsia, or postpartum anemia. Most women saw their doctor at a minimum 11 times, at most about 14 times. For the most part, the investigators found that women appreciated more frequent visits, if only for the reassurance that everything was going along smoothly.

The real question would be how often do things go wrong when doctors and nurses are not around? Without reliable statistics, no one really knows. But even if the risk is minuscule, isn't that enough of a risk to want expert help nearby?

In the 1950s, Pat Carter made a few headlines not because anyone considered her a trailblazer but because they thought she was weird. Are the unassisted birthing women today any different? They may have garnered a bigger following, thanks to cyberspace.

They may land deals with television shows, thanks to daytime talk shows. But have they grown beyond the freak-show phenomenon? Will they change birth practices? Freebirthers would say that they are not revolutionizing birth so much as encouraging women to deliver their babies in the most peaceful way possible, or at the very least to think about why they are making certain decisions. Freebirthers insist they are not missionaries for their way of birth. They just lay out the facts to educate women about birth choices. They say they are trying to convince women that even if they do not choose the unassisted route, they should be happy with who is doing the assisting.

12

Womb with a View

The waiting room at View a Miracle, a photo studio for fetuses, was packed with pregnant women, husbands, toddlers, and expectant grandparents. View a Miracle is located in the Lion's Plaza, a strip mall in New Jersey, about an hour's drive south of New York City. For $95 to $200, depending on the package you choose, you can get a DVD and a CD with a selection of three-dimensional images of your minus-2-month-old. This is not a medical exam. The sonographers are not doctors. They do not diagnose. They are not allowed. "You spend so much on the pregnancy and baby, what's a few hundred dollars more," said a pregnant customer in the waiting room. Why not?

It's really not a waiting room in the sense of a doctor's waiting room. It's more a boutique. Until it's your turn, you can buy baby clothes and frames for the fetal photos. There were so many choices. So many puns. One frame had teddy bears with the caption "I can *bearly* Wait to See." A frame with a toy truck border said "Under Construction."

The photo studio was tucked in the back. It had two television sets, one for the mother and a second for her guests. There is seating along the wall for 25. Video streaming is available for an extra fee. (Shortly after the visit, a teacher said she "freaked out" when a 3-D photo of her friend's fetus popped up on her cell phone.) The room was not quite living room cozy, but it wasn't doctor office staid either.

"Some people don't really want to see their babies before they are born," said Kayla Gipson, one of the sonographers. The first clients that afternoon were a 30-something husband and wife. They did not bring guests. The woman got into ready position on the examining table—reclining with her head propped up so she could see the TV, her shirt lifted above her belly so the sonographer could rub her abdomen with the gel-coated ultrasound wand. Soft muzak played. It was a medley of Brahms's "Lullaby," "London Bridge Is Falling Down," followed by—can you guess?—Roberta Flack's "First Time Ever I Saw Your Face."

In the advertisements, the images seem like a shadowy image of a baby in a water-filled ziplock bag. Shadowy but cute. In real life, it is eerie. The unborn babies look like orange glow-in-the-dark ghosts. Picture a sunburned Casper the Friendly Ghost pressing his face against a window. Cute to some. Creepy to most—especially if it's your kid. Sometimes chunks of the head are missing. That's because, depending on where you put the probe, the placenta covers the body and appears as a black blob. It makes the face look like a mouse has nibbled a chunk out of it. It has nothing to do with the machine or the sonographer, but how the baby happens to be posing in relation to the placenta.

The sonographers said ultrasound images are reassuring. The first dad looked frightened and nauseated staring at his fetus that had a mouse-eaten head and smushed nose. He did not say a word, but the sonographer sensed his concerns. She explained that all fetuses have wide, flat noses because the cartilage has not formed yet. (Cartilage stiffens after birth.) She said she would try to get a shot without the placenta shadows. No guarantees, of course.

The next couple were African-Americans. A nice head shot of their little baby girl fetus appeared right away. The sonographer pointed to the wide nose (the baby looked remarkably similar to the one before) and cooed, "Oh, look, she's got your nose." So much for the soft cartilage. As the sonographer moved the probe, the face disintegrated and then popped back on the screen.

View a Miracle is not unique. Entertainment fetal photo shops are popping up all over America. There is Womb with a View, Womb's Window, Sneak Peek Ultrasound, Baby Waves, Peek-a-Boo, and so many more. Some shops, such as View a Miracle, are owned by fetal sonographers with years of experience—the sonographers at View a Miracle work for doctors during the day and have their own business in the evenings. Other shops are part of national franchises that provide weekend training courses. The photographers are just that, photographers. They can tell a nose from an elbow, but they are not trained to pick up heart defects or other fetal anomalies, nor are they allowed to divulge medical information aside from saying boy or girl. Needless to say, doctors—many of whom offer their own entertainment fetal photos to patients—are not amused. Doctors say they are concerned about excessive exposure to high-frequency sound waves for anything other than diag-nostic purposes. Shop owners say that they have more time than doctors to ooh and aah with each patient, spending upward of an hour for each photo session.

"a craze unlike any that had come before"

The notion of prying into the once private lives of fetuses began long before ultrasound. It began when Wilhelm Roentgen discovered what he called "a new kind of light" that he dubbed X-rays, as in X for the unknown. As Bettyann Holtzman Kevles tells it in *Naked to the Bone*, Roentgen irradiated his wife Bertha's hand for 15 minutes, sent the image of her bony hand to a science journal, and launched an overnight international sensation. It was "a craze unlike any that had come before . . . broad, swift and undiluted by fear," Kevles wrote. The new photo machine was irresistible. Doctors built their own X-ray image makers and filmed everything—burning, maiming, and killing themselves with overdoses of radiation along the way. By the 1930s, X-rays were a routine part of the prenatal exam. Doctors used them to make sure the fetal thigh bone was connected to the knee bone, and so on. They used them to measure the fetal head and the mother's pelvis to see if the baby could squeeze through. If there were any doubts, they prepared for a C-section. One doctor blamed X-rays and faulty measurements for prompting too many surgical births. (The C-section rate was in the single digits then.)

Few folks worried, until, that is, an extensive survey conducted by an Oxford team of investigators led by Dr. Alice Stewart linked fetal X-ray exposure to childhood leukemia. She compared thousands of mothers who lost a child to leukemia with thousands of mothers with healthy toddlers, asking them everything about their pregnancy and shortly thereafter: what they ate, what they did, how often they were in cars and buses. The questions about pregnancy were a unique addition because most researchers gave little thought to fetal exposure. "We could see it quite early on, from the first thirty-five pairs, yes was turning up three times for every dead child to once for every live child, for the question, 'had you had an obstetric X-ray?'" Stewart told Gayle Greene, her biographer. "They were like two peas in a pod, the living and the dead, they were alike in all respects except on that score. And the dose was very small, very brief, a single diagnostic X-ray, a tiny fraction of the radiation exposure considered safe, and it wasn't repeated. It was enough to double the risk of an early cancer death."

The preliminary results were published in a 1956 issue of the medical journal *Lancet*. Stewart told Greene she was excited because she knew she was on the verge of shaking up the medical community. Few obstetricians shared her glee. They questioned the findings and attacked the study because it was based on a survey, a soft science. Here was a young doctor, a woman no less, telling them the standard of care was killing patients.

Stewart reviewed and reanalyzed her data. The same findings emerged and reemerged. Other researchers did their own studies. Their findings confirmed and reconfirmed the Oxford study. And yet, doctors continued to do the fetal X-ray exam for another 20 years. It's "like anything else, there was still a transition time," Dr. Barry Goldberg said recently. Goldberg, who is now director of the division of ultrasound at the Jefferson Ultrasound Institute in Philadelphia, was one of the few doctors who fought for a national ban on fetal X-rays after the Stewart results were revealed. "X-rays were still being done because doctors were used to them," he said. Insurance companies reimbursed for a fetal X-ray, but they were not yet paying for the brand new ultrasound images. The end for fetal X-rays came about not so much because doctors finally accepted the dangers, he said, but because insurance companies began to reimburse for the ultrasound exam.

"There is not much difference between a fetus in utero and a submarine at sea. It is simply a question of refinement."

Ultrasound machines were invented to spot German submarines during World War I. Scientists realized that they could control the beam of sound if they used ultra high-frequency energy, faster than 20,000 waves per second, so fast it cannot be heard. Hence the name *ultrasound.* Military officers used the acronym SONAR, for sound navigation and ranging. Hence the name *sonography.*

Years later and another world war later, Ian Donald, a Scottish obstetrician, figured that "there is not much difference between a fetus in utero and a submarine at sea. It is simply a question of refinement." He aimed an ultrasound machine at pregnant bellies and launched a new era of spying on squiggling unborn babies.

Donald found his first embryo sighting accidentally. He could easily have missed the discovery had it not been for his keen eye and wickedly smart insight. He was examining a woman who had several miscarriages and used his new ultrasound machine to search for fibroids. (The hospital did not have a machine so Donald borrowed one from a local factory owner who used ultrasound to find flaws in metal.) On examination, he saw a white circle and knew it was the embryonic sac. "I went to New York and I addressed the New York Obstetrical Society and I said, 'just look at this!' and they looked totally unbelieving. I said this ring's a gestational sac—that's a pregnancy, and my God, it was. I delivered that baby with my own lily-white hands at term—a perfectly healthy baby, her first."

Donald began to collect embryo images. One of his students, who apparently shared his zeal, was on an airplane when she started to miscarry and saved the excreted tissue for ultrasound examination. Donald described the gory scenario to Ann Oakley, a University of London sociologist:

> She goes to the loo and passes something. Now this is the bit I like her so much for, bless her. Instead of turning on all that blue stuff that goes down the loo, she picked it out and put it in a bottle, if you please. She then got to her destination, not only did she have it examined in the (pathology)

lab, she had it photographed under water and sent me the photograph. Then she had it examined by the cytogeneticist and it would have been a mongol. And it's in my book: "By kind permission of the mother, herself a gynaecologist."

In the early days, ultrasound produced spikes and dips on a screen, more like looking at an echocardiogram image than a fetal photo. The images were meaningless to anyone except those who were trained to discern what each blip meant. The trick was to figure out how to turn the stripes into user-friendly images. The images got much better, thanks to Douglas Howry, at the University of Colorado Medical Center; John Wild, at the University of Minnesota; and George Ludwig, at the University of Pennsylvania.

By the late 1960s, ultrasound was becoming increasingly popular, but not nearly as routine as it is today. Donald, the most fervent investigator and one of ultrasound's greatest fans, worried about the misuse and overuse of ultrasound. In one of his earliest papers, he warned that the "utmost vigilance must be maintained because in the future more sophisticated and more powerful machines may one day introduce hazard that is at present hardly foreseen. . . . We must not forget that it took nearly half a century for the damaging effects of X-rays upon the fetus in utero to come to light."

"loophole big enough to drive a truck through"

In the 1980s, ultrasound was a routine prenatal exam; the number of photos depended on your doctor and your insurance company. For the most part, everyone was excited. What was not to love? Just a generation earlier, doctors listened for that reassuring sound of the fetal heartbeat with a stethoscope and felt a woman's abdomen to monitor how the fetus was growing. It was educated guesswork

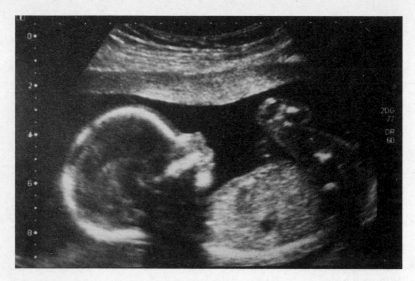

Two-dimensional fetal ultrasound at 22 weeks. istock.com.

really, imagining what the baby looked like from hearing and feeling. Ultrasound images in the 1980s were static on a television screen to the untrained eye. Expectant parents squinted, trying to see something humanlike when their doctor pointed to a kicking leg, a wriggling hand, or the face. The session's climax was the heart. Even if you couldn't tell a big toe from a nose, it was pure glee to see the fetal heart, an ever-so-tiny wisp of white chalk flickering back and forth. That first glance made everything real to the expectant mother. Yes, I am really carrying a baby. There is life inside. Suddenly, one stroke of the gel-coated ultrasound wand transformed what had been a few missed periods and swollen breasts into something real. And alive.

Yet amid the excitement, anxiety simmered. Health activists, women who were aware of the DES fiasco, the thalidomide travesty, and the dangers of X-rays, worried about lurking dangers in ultrasound. Insurance companies wanted proof they were paying for a medically necessary exam. In February 1984, in response to the brewing concerns, the nation's leading childbirth experts convened at the National Institutes of Health in Bethesda for two days to figure out what to do with fetal ultrasound. The issue was not whether ultrasound worked. Everyone knew it did. The machines were getting better all the time. By this time they produced fuzzy two-dimensional images with snazzy take-home pictures. That part was unequivocal. And yet, there were two lingering questions: Was it worth it to scan every single pregnant woman? And were there lurking dangers?

The group, funded by the Food and Drug Administration and the National Institutes of Health, had a mandate. They had been told to review all available data and come up with a consensus. Their guidelines, announced that February, would signal to women whether ultrasounds were a good thing to do during pregnancy. What the doctors decided would influence whether insurance companies would pay, and if so, how many ultrasounds per pregnancy.

From the get-go, the doctors realized that regardless of their decision, ultrasound, like all technology, was destined to go in one direction: more portable and more popular. And yet they knew they should provide advice. Their findings would not be law, but expert opinion. The time was right for such an appraisal. Advances in computer technology were improving ultrasound machines dramatically, making interpretations of the images much easier. Doctors, in increasing numbers, wanted to buy them. Hospital administrators realized that the machines had evolved from costly research gadgets to moneymakers.

Beth Shearer, who testified before the panel, said she was frightened that doctors were getting carried away with potentially dangerous technology, making women feel as if they were "no longer capable of safely carrying birthing babies without a lot of machines, tests, and interventions." She and other women came armed with data suggesting that ultrasound was dangerous. The doctors had already spent the better part of the year assessing the same studies and were not convinced of the hazards. One study, for instance, aimed ultrasound at a cluster of cells in a petri dish and the chromosomes mutated.

Dr. Frederic Frigoletto, a Harvard professor who chaired the 1984 panel, said that the ultrasound doses to the cells were much higher and the exposure much longer compared with the typical routine exam given to women. Besides, cells sometimes mutate in the laboratory dish when nothing is done, so you cannot blame ultrasound. In another study, trumpeted by the women and trashed by the experts, Japanese scientists aimed ultrasound at a group of pregnant mice. Some mice were born with missing limbs. As it turns out, the study was done on a group of mice prone to limb defects. Again, you cannot blame ultrasound.

And yet despite their skepticism about the studies, the doctors were not convinced that ultrasound was completely safe. "We can never dismiss the fact that there is some possible risk," said Dr. Frigoletto. He has a hunch that the commotion at that meeting prompted manufacturers to monitor more carefully the power of their machines.

The panel of experts declared that prenatal ultrasound should *not* be used routinely. Colleagues were stunned. Though the panel acknowledged studies suggesting that ultrasound was not dangerous, they said there was not enough proof of safety. And yet, they did generate at least a dozen reasons to warrant an ultrasound exam—reasons when the benefits of seeing the fetus early would outweigh a potentially small risk.

No one was happy. Doctors planned on offering women routine ultrasounds regardless, banking that insurance companies would continue to reimburse. (Since the late 1970s, most insurers had paid for at least one prenatal exam.) Women's activists were mad, too. Doris Haire, who also testified that day, called the long list of valid medical reasons a "loophole big enough to drive a truck through." If you wanted a fetal photo, all you had to do was insist that you could not remember when you conceived the baby, and you had an automatic medical reason.

Few doctors forecasted the marketing bonanza of the twenty-first century that turned the 1984 ultrasound controversy into a quaint tale. On eBay, you can buy a fetal ultrasound scrapbook that has a poem on the first page that begins, "I loved you from the start." It comes with "I loved you from the start." Care Bears stickers. One of them is a pregnant rabbit eating pickles and ice cream with the caption "ultrasound cravings." If you're really enthusiastic, you can buy your very own machine for personal use or to make some extra money. Machines range from $15,000 to $200,000.

The marketing of ultrasound like photo booths in malls rankles physicians. In 2005, fetal ultrasound specialists launched the Keepsake Ultrasound Task Force under the umbrella of the American Institute of Ultrasound in Medicine. Participants worried that women will get overdosed with ultrasound or get false reassurance from the photographs. Though safe for diagnosis, "ultrasound energy has the potential to produce biological effects," the committee concluded.

The FDA is angry too. It opposes the use of medical devises for nonmedical uses. But it has not shut down the centers. A few doctors suspect that the FDA's enforcement is lax because if they shut down the entertainment centers, they'll have to take away the ultrasound machines from the antiabortion centers, too. Antiabortionists have tried, unsuccessfully so far, to pass laws requiring that every pregnant teenager watch a three-dimensional ultrasound of her fetus before having an abortion.

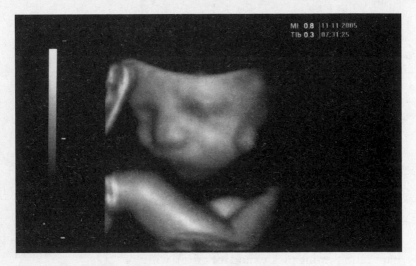

Three-dimensional fetal ultrasound at 21 weeks. istock.com.

   Ultrasound is energy. High doses are used to heat and heal muscle injuries. It is used in other countries, but not the United States,
as an alternative treatment for cancer. If it's therapeutic—or if
it's changing muscle physiology somehow—there is a chance that
it could affect the baby, particularly at excessive doses for a long
time. Doctors say that the low dose used for a few exams during
pregnancy are safe, that the benefits outweigh any potential minimal risk. They worry about excessive doses for long periods. Any
risk, no matter how minuscule, outweighs the benefits when there
is no medical reason for the exam, say doctors.
   Advocates insist the photo session helps mothers bond with
their babies. They point to studies that seem to defend their case.
Dr. Joshua Copel, a professor of obstetrics and gynecology at
Yale University and president of the American Institute for Ultrasound in Medicine, advocates three-dimensional ultrasound for
medical uses only. "My own take," he said, "is that for several
million years, mammals have bonded to their young without ultra-

sound and we have done reasonably well as a species based on our mother's nurturing us until we can take care of ourselves. There is no reason to think we need something additional to feel the right things towards our babies. I say to women, if you want it do it, but the relationship you form with your child is how you respond for the next 18 years. I'm not convinced 3-D ultrasound has anything to do with it."

Ultrasounds will inevitably become clearer and will be used more frequently as part of the prenatal exams. On the bright side, couples can go home from their doctors with snapshots, reassured that everything is okay and with some concrete evidence that their baby—at least physically—seems to be growing according to the books. The problems arise as our technology advances faster than the ability to interpret it. What happens when the doctor spots a mark on the baby? Could it be a defect in the image itself or does it signal greater troubles? What happens if there is something that does not quite look like every other baby, but no one is sure what to do. In 1996, *New York Times* science writer Natalie Angier wrote about the harrowing ordeal when she and her husband were told that their 20-week fetus may have clubfoot. They spent the rest of the pregnancy preparing for the worst, and then the baby was born completely healthy without any signs of clubfoot.

When ultrasound works well, it picks up defects that can sometimes be corrected before birth, sometimes shortly after. Ultrasonographers continue to defend the use, pointing to all the great things that the imaging machines have done to help women. But bioethicists worry. They are concerned about the growing numbers of women, like Angier, who are told that something may be terribly wrong when the baby is perfectly fine. (Doctors did tell Angier that her baby may have had clubfoot that corrected itself before birth. Did seeing the foot help anything?) They also fear that images deemed healthy can be falsely reassuring. Dr. Sheryl

Burt Ruzek, a professor of health education at Temple University, told Angier that ultrasounds, particularly when a pregnant woman has several, have "created false expectations that by having repeated screenings we can improve the likelihood of a good outcome of pregnancy. The search for the perfect child is making women very anxious about reproducing."

The advent of sophisticated user-friendly fetal ultrasound speaks, in many ways, to our image-oriented society. We need a film clip (something to put on YouTube, perhaps?) to show the world and to prove to ourselves that what we think is happening is real and meaningful. And yet, fetal ultrasonography has made a much more profound impact on pregnancy, one that had not been foreseen by its pioneers. In the beginning, the pregnant woman was housing an unknown, invisible creature. She alone was the patient. The sophisticated microscopes and imaging tools have given the fetus its own identity. There is now a full-fledged medical field of fetology along with a batch of legal and ethical debates about the care of the unborn. As Kevles points out in *Naked to the Bone*, "Ultrasonic images have contributed as perhaps no other imaging technology to polarizing attitudes about the personhood of fetuses, embryos, even unfertilized eggs."

Part Five

13

Sperm Shopping

The most gorgeous guy is on a California sperm bank Web site. He has chiseled features, wavy blond hair, and a toothy smile. He's wearing a cap and gown; presumably, he graduated from somewhere. He'd be the perfect specimen if he were donor 3536. Most donors are anonymous, so you cannot connect photos to numbers. For that matter, shoppers have no way of knowing whether the sexy cover boy even left a deposit at the bank. The point of the photo is to whet your appetite for the sorts of men who store assets, not to claim that this guy really did.

In a handwritten essay scanned online, donor 3536 wrote: "I am a funny and easy going type of person. I'm also a hard worker and always striving to make my dreams come true." His message to his potential semen recipient: "You should enjoy life and do everything that you think is possible. It is important to stay focused on your dreams, but it is also important to sample all the offerings life has to offer." His hobbies included good things, such as drawing, photography, travel, and biking, but he also added video games. Would the kid be addicted to Play Station? That was a scary thought.*

For a fee, you can get more information about 3536's likes and dislikes. At least shoppers can assume he's decent looking, because ugly men are rejected. Even if you're not in the mood to buy sperm, it's still fun to window-shop. With so many Web sites, sperm shopping is easily accessible to anyone who wants to try to understand this relatively new and booming business and to get an idea of what so many women have gone through to experience pregnancy and have a baby that is genetically linked to them.†

From the buyer's perspective, it appears fun at first glance. You are the queen bee, choosing your little worker bee. But the more you seriously delve into this modern game of house, making believe you really want a donor dad, reality will hit. Sperm shopping makes dating seem simple. When you're in the market for a man, you have a few drinks, and with any luck, you fall in love and in time make babies. Then you take the bad with the good. When you have to pick a sperm, you analyze the pickings from a Web site; the skinny vial is shipped overnight to a fertility specialist; and then, if all goes well, you get pregnant and go to your next specialist, the obstetrician. To get your man—your man's sperm, that is—you deal with a lot of middlemen. You are not picking a spouse. You are trying to create your dream child.

That means that when it comes to sperm, you want perfection. If the end product, your child, does not come out as you imagined, will you be disappointed? Will you blame yourself? File a lawsuit? All of this assumes that you believe in genetic destiny with less

---

* I am not sure what the advantage is of a handwritten essay—except perhaps you could hire a handwriting specialist who could provide even more clues to the psychological state of your potential donor, which some banks say they offer.

† Artificial insemination is not new, but turning what had been a secretive medical treatment into a moneymaking business is.

concern for the nurture part of the equation. You're thinking that if your donor likes surfing and following his dreams, it must be inscribed in a dominant gene that is guaranteed to pass along to your kid.

Contrary to the situation when you're looking for a man— all single women complain there aren't enough decent ones around—when you browse for sperm, you realize the pickings are mind-boggling. Each sperm bank—there are upward of some 20 for-profit outfits—has their own Web site with spreadsheets of sperm donor descriptions. There are also Web sites to help you sort through the Web sites. Later on, you can find Web sites to help your donor-made child hunt down the anonymous donor dad or a half-sibling.

The doctors who own the sperm banks are more bankers than doctors. They are not curing anything or even working with infertile couples. They store and sell assets. They broker deals between sellers and buyers. What had been a hush-hush medical procedure of doctors procuring batches of fresh sperm for infertile couples has blossomed into a global enterprise with sperm sales skyrocketing as single women and lesbian couples join the hunt. (In the old days, banks shunned lesbians and singles. Now they court them.) One bank said they ship about 30,000 sperm ampoules every year to every state in the United States and around the world.

Sperm banking is a lucrative business with few regulations. According to a *Wall Street Journal* article published in 2000, global sperm exports totaled somewhere between $50 million and $100 million. In 2005, decades after many banks had been in business, the U.S. Food and Drug Administration implemented disease prevention regulations. According to the FDA, all frozen sperm must be tested for specific diseases, such as HIV; and sperm donors must be screened for infectious diseases. To get a seal of approval from the American Association of Tissue Banks (it's not required), sperm must be tested for the genes linked to Tay-Sachs disease, thalassemia, sickle-cell trait, and/or cystic fibrosis if the family history or ethnic background indicates that the donor is at risk. In addition, every donor must provide at least three generations of family history, to be evaluated by a geneticist. This last stipulation assumes that the donor knows his family history and is telling the truth. Also, anonymous donor sperm must be quarantined for six months and retested prior to release, according to the AATB. Of the 20 or so commercial sperm banks (there are many other banks that store husband sperm only for IVF but do not sell anonymous sperm), only about 9 are accredited.

That may seem like a lot of rules, but there are so many things that are not included in the regulations. The last time anyone tallied the number of ampoules of sperm sold or babies born was in 1988 by the now-defunct Office of Technology Assessment. It estimated that about 30,000 babies were born from donor sperm between 1986 and 1987. Since then, no one has kept track of how much sperm is bought and sold nationally and how many babies are from donor sperm or, most importantly, how many babies are born from each donor. There is no way to know whether your sperm donor baby has one or 50 half-siblings. There is no way to know if the babies born from a particular donor suffer from the same genetically linked disease because no one is keeping track. As always, technology and business are a few paces ahead of the legal and ethical debates, which are addressed as crises occur. Already parents have sued for children born with genetic diseases.* Tens

---

* In 1999, Ronald and Diane Johnson sued California Cryobank because their daughter, Brittany, born in April 1989, suffered from a serious genetic kidney disease, a disease linked to a dominant gene not in Diane's family. The Johnsons sued for punitive damages as well as fraud. In 1999, the court ruled that Brittany could not recover damages. The court found that the docket of information the family got was different from the profile information stored in the bank. The law stated that a child can recover if her doctor breached standard of care and caused

of thousands of children are hunting for donor dads, despite a promise of anonymity. State courts are deciding whether a lesbian can claim custody rights to a child she parented yet was born to her partner with donor sperm. Outside of the courts, parents are grappling with the inevitable and vexing issues of whether or when to tell children how they were born and whether to help or encourage children to find their donors. Do you call them donor dads or tissue donors? One sperm donor says he calls his offspring (he has met two) his biokids. Semantics shape the argument.

In 1867, J. Marion Sims, the same doctor who perfected the vaginal tear repair by operating repeatedly on slaves, appalled colleagues by predicting that there will be a day when scientists will figure out when women are most fertile and cure infertility by placing sperm into vaginas. As he put it, "This mechanical fertilization might become exact enough to depend upon it in such cases as would otherwise be impracticable." He was thinking of using sperm from the husband. I seriously doubt that he was thinking about donor sperm, and he probably never fathomed a lucrative trade.*

---

pain. It was a matter of legal decision whether the bankers were caregivers. They weren't. They sold a commodity. The doctors did not cause an abnormality in the baby. The disease was caused by genes in the sperm the parents chose to purchase. L. Andrews, *The Clone Age: Adventures in the New World of Reproductive Technologies* (New York: Holt, 1999). The court's opinion can be found at "Natural Selection in Family Law," 5.1 Bad Sperm, at www.biojuris.com/natural/5-1-0.html.

* The sperm trade is okay for humans but not thoroughbred horses. Thoroughbreds, according to the American Jockey Club, have to be made the old-fashioned way. According to Section V, rule 1-D, of the stud book, "to be eligible for registration, a foal must be the result of a stallion's Breeding with a broodmare (which is the physical mounting of the broodmare by a stallion with intromission of the penis and ejaculation of semen into the reproductive tract). . . . Any foal resulting from or produced by the processes of Artificial Insemination, Embryo Transfer or Transplant, Cloning or any other form of genetic manipulation . . . shall not be eligible for registration." I found this out because when I mentioned sperm banking to a friend, he wondered whether anyone saved the sperm of the 2006

Sperm are wonderfully complicated. Unlike eggs, which are made before birth and distributed one at a time until supplies run dry, sperm is constantly manufactured from a finely tuned internal factory. The machine starts cranking at puberty. Counts vary dramatically among species. According to one intriguing theory, the most promiscuous mammals have

"By the time you get to gorillas, it's a disaster."

the highest counts. The thinking is that if females have sex with several mates, males need high counts to override the competition and ensure offspring. Dr. Sherman Silber, microsurgeon and sperm expert at St. Luke's Hospital in St. Louis, believes in the promiscuity–sperm count theory. He points to bulls and chimpanzees, whose libertine females sleep with any passing male. These virile males spew about a billion sperm per ejaculate, one of the highest counts in the animal kingdom. The gander mates for life with his loyal goose and has small testicles and a low sperm count. "By the time you get to gorillas, it's a disaster," said Silber. They're monogamous and release fewer than a million sperm per ejaculate. Silber said human counts have declined as we have become a more monogamous society. Not everyone agrees. Some scientists blame lower sperm count on environmental hazards. Others do not believe there has been a decline.*

On average, men release about 200 million sperm per ejacu-

Kentucky Derby champion, Barbero, who hurt his leg and was euthanized. The answer is no. He and his sperm were cremated.

* The promiscuity theory, I should emphasize, refers to species—not individuals. One scientist told me there have been reports of low sperm counts among monogamous men, prompting some scientists to claim that there may be some validity to the theory. I don't believe it. They just say it to make the wives of low-sperm-count men feel better.

late and then replenish the well. It takes about 64 days to make a sperm. There are always sperm in every stage of production. Despite the big numbers, few sperm have what it takes to get to the egg, and even fewer sperm have the wherewithal to get inside. From a woman's perspective, a successful marriage of sperm and egg is like dating on a microscopic scale. There are so many losers out there. Most eligible sperm are inept. Some straggle. Some go the wrong way. They're all slow. Sperm swim about 30 micrometers per second, which means it would take a sperm about 10 minutes to swim across the period at the end of this sentence. Fortunately, the force of ejaculation propels most sperm across the starting line, giving them a much-needed boost. In the case of artificial insemination, they start at the finish line. Intense muscle contractions of the vagina facilitate the journey. Getting from one end of the birth canal to the other can take anywhere from 5 minutes to an hour. Whereas invertebrate sperm must keep swimming or die, human sperm sometimes wade. They tread in place and then rev up as they near the egg.

Swimming to the egg is only half the battle. Once they are there, each sperm is one among thousands knocking at the door. Only one will give the magic kiss that opens the gate and shuts the others out. Specific enzymes at the head of the sperm digest an outer portion of the egg, permitting entrance. It is called the acrosomal reaction.

Sperm specialists cannot tell you what distinguishes the fertilizers from the losers. One sperm banker told me that his team checks for the speedy swimmers and disqualifies the ones that go round and round. But swimming crooked may not be a bad thing after all. Recently emerging evidence suggests that sperm have a receptor called hOR_{17-4} that guides it to the destination. Each sperm, in essence, is endowed with a global positioning system that navigates the way through the birth canal. That means that zigzagging is intentional, not reckless driving.* Jeffery A. Riffell, a neuroscientist at the University of Arizona and one of the researchers, likes to think of it as the sperm having a nose, sniffing its way to the egg. "The interesting thing," he says, "is that the cells that can smell, that can chemosense, are the ones that undergo the acrosomal reaction. There is this very strong correlation between the ability to smell and ability to fertilize. It's really weird that these single cells have all this machinery. It's an incredible selective force operating." Dr. Riffell also suspects that some sperm sense heat and can detect the slightly higher temperature surrounding the egg. He says that the receptor is only one among a slew of receptors, which makes sperm more complicated noses. No one knows what these receptors are doing.

Even with the extra perks, the journey is not smooth sailing. It's not as if sperm are fish racing through a tunnel. It's more like tadpoles wading through tumultuous muck. The birth canal contains a thick, viscous fluid with an undercurrent that pushes sperm backward, forward, and sideways. The chance of fertilizing an egg is not simply a question of speed but speed in relation to vagina tumult. As Riffell and his colleague Richard K. Zimmer, of the University of California, Los Angeles, explained in the *Jour-*

---

* Like many scientific "discoveries," there was always someone earlier saying the same thing. In the late 1800s, J. Marion Sims noted that sperm seem to wiggle around as if they are looking for something. And then in 1939, Dr. Frances Seymour commented that doctors should not judge sperm by looks. Her studies found that "the fertility index of spermatozoa may not be judged by conformity of the count per cubic centimeter, by ascertainable motility or by any other physical factors diagnosable by ordinary microscopic examination." These were the days before doctors were talking about receptors, but she was onto something that today's researchers seem to be confirming. F. I. Seymour, "Sterile Motile Spermatozoa Proved by Clinical Experimentation," *Journal of the American Medical Association* 112, no. 18 (1939): 1818.

nal of Experimental Biology, it's all about the F_{swim}/F_{shear} ratio, which sounds to me like a formula for judging potential Olympians. When the speed of the sperm outdid the speed of the fluid, the sperm were able to fertilize the eggs with greater success. Sperm are swimming against a tidal wave. If they catch the wave at the right moment, it may propel them to the egg. But they can also get shot backward.

The evaluation process is problematic. Riffell wonders whether sperm act differently when viewed with a microscope: "In terms of whether sperm swim straight or turn has to do with how they are positioned on your microscope slide. The ones in circles are just trying to swim normally, but they are trapped. The ones that are straight have not come into contact with the surface. I guess what I'm saying is that there may be some artifact involved." In other words, we are judging our sperm under false pretenses.

In the fourteenth century, Arnold of Villanova, a Spanish doctor, tried to create a baby by putting semen into a womb-shaped vase. He thought, as so many others did, that sperm contained miniature humans that blossomed inside the womb. It didn't work. Another doctor tried again a hundred years later. It still didn't work.

"Better let ancient families become extinct than keep up succession by such means."

Doctors started putting sperm into vaginas when they figured out that women were a necessary part of the fertilization process and not simply the soil from which babies grew. In 1785, famed British surgeon John Hunter helped impregnate the wife of a wealthy businessman. The husband ejaculated into a jar and Dr. Hunter used a quill to scoop up sperm and put it into the wife. For the most part, doctors used husbands' sperm, and for the most part, it was a highly secretive enterprise.

In 1867, J. Marion Sims's textbook *Clinical Notes on Uterine Sur-*

gery with Special Reference to the Sterile Condition described in precise detail what doctors had been doing for years. He called it "artificial fructification." Colleagues were furious. The public was flabbergasted. Dr. Sims said he used a humungous glass syringe to suck up semen in the woman's vagina, shoved the instrument further back, and squirted the semen closer to her uterus. Sometimes he knocked her out with ether first. This procedure occurred immediately after intercourse. You can't help but wonder where Dr. Sims was the whole time. Waiting outside the bedroom? And then there's the wife lying there naked with legs spread. If I were her, a young married woman in Victorian America, I'd ask for ether, too.*

The *British Medical Journal* considered Sims's book too tasteless to review. The *Philadelphia Medical and Surgical Reporter* printed a quote from Hamlet to poke fun of the whole thing: "To What base uses may we come, Horatio?" The *Medial Times and Gazette* of London mocked Sims for "dabbling in the vagina." They said that couples were better off without children than resorting to such inappropriate means. "Many things have been described on paper which have hitherto been veiled in Professional silence, even if they entered the imagination of professional men. We can but express an unfeigned regret that Dr. Marion Sims has thought proper to found an odious style of practice on such methods. . . . Better let ancient families become extinct than keep up the succession by such means." In a subsequent issue, the *Gazette* published a poem submitted by a reader. It was called "Ode to Dr. Marion Sims."

---

* According to F. N. L. Poynter, MD, former director of England's Wellcome Trust, who wrote about the history of artificial insemination, Sims's account angered Americans and English, but the French used it as a starting point for more public discussion. Artificial insemination was used more widely in France.

Say, what is man? An atom at the first,
Waiting its nuptuial atom in the womb;
Too oft, alas, by fate untimely curst,
In place of fostering home, to find a tomb.
Grieved at the thought, a tear thine eye bedims,
Great son of Aesculapius, Marion Sims

Swift to thine aid inventive genius brings
Persuasive tent or glistering hystrotome:
With these the obstructed portal open flings,
And guides the struggling sperm'tozoon home.
Thus may the wished-for union perfect be,
The mystery no human eye can see.

Sims, should these fail, thou still wilt cherish hope
To find some other cause that breeds the ill;
With learned digit, searching microscope,
Or peering speculum, exploring still:—
Nay, wizard-like, ethereal sleep wilt shed
To win they point, e'en oe'r nuptial bed.

There was no uproar over his cruder experiments, when he invited guests to watch as he stitched and restitched slaves' vaginas without anesthesia. But when Sims went into the bedroom of married white women to help them get pregnant, doctors were enraged. It was not just the experiment but Sims's newfangled notions about making babies that enraged the medical community. Women who got pregnant while knocked out with ether debunked the long-held notion that a couple needed simultaneous orgasms to make babies. He wrote that if great sex were necessary to make babies, humans would be fossils by now. "It matters not how awk-

ward and unsatisfactory the act of coition may be performed,"
he wrote, "so that semen with the proper fructifying principle be
placed in the vagina at the right moment."

He also had an abysmal success rate. One woman, whom he
considered a success, got pregnant after ten tries and then miscarried four months later. The problem was that doctors did not
know when women were ovulating. Oftentimes he put the sperm
in during the menstrual cycle.

The first published account of donor sperm was in 1909,
describing an insemination that had occurred 25 years earlier.
According to Dr. Addison Davis Hart, his professor, Dr. William
Pancoast, selected the handsomest medical student to donate a
fresh sample. Pancoast never told the wife
that she was getting a stranger's sperm. She
thought he was testing a new fertility treatment. Some historians think that Hart was
the donor—hence the reason he wrote the
article and seemed to know the details of
the deed and added the bit about the sperm
donor being the best-looking medical student.
Apparently, Pancoast never published because
he wanted to keep the procedure a secret.

"he has an
intelligence
quotient of
140" and
therefore
"children . . .
would be
highly
desirable"

The sperm trade has always been about
selecting the best and the brightest, a laboratory approach to the blossoming eugenics movement, and
it's always had its cowboys and critics. Frances Seymour, a New
York City physician, used semen from donors with high IQs and
in 1935 agreed to work with a university professor and his wife
because "he has an intelligence quotient of 140," which throws
him into the genius group, and therefore "children . . . would be
highly desirable." The following year, Fred Hogue, writing in the
Los Angeles Times, called Seymour's experiments and those of oth-

ers who tinker with donor sperm "repulsive." Or as he put it, "17 children born out of wedlock! Who will be responsible for the support of these children? What will be the social position of the mothers of the 17 illegitimates? Will these children be permitted to be born, or will some form of abortion—purely in the interest of science—be used to do away with them?" Even into the twentieth century, critics called babies born of artificial insemination "artificial bastards." It's been said that Aldous Huxley's *Brave New World*, published in 1932, was more critique than prophesy.

For years, the secretive sperm business generated critics who worried about the ethical and legal ramifications. In 1947, New York City passed a health code that only doctors could collect, sell, or give away sperm. In 1954, an Illinois court ruled that donor sperm was adultery. In 1981, the Wisconsin legislature attempted to pass laws prohibiting doctors from performing artificial insemination on any woman likely to become a dependent on government aid. The governor vetoed the bill.

The first recorded birth from frozen sperm was in 1953. Historians Cynthia Daniels and Janet Golden, in their comprehensive article about the history of the sperm business in the *Journal of Social History*, say that the ability to freeze plus overnight delivery propelled the sperm industry from a local enterprise to a global affair. What began as 10 banks in the United States in 1969, they wrote, exploded to 135 banks by the late 1980s. It was a wildly unregulated industry. For the most part, clients were married heterosexual couples. A few banks catered to lesbian couples. One bank claimed to store genius sperm.

But a few cases of HIV detected in donated sperm dramatically changed the industry. The expense of recruiting and screening donors for the AIDS virus and other diseases drove out the small players. As Daniels and Golden see it, today's industry plays by the same turn-of-the-twentieth-century eugenics ethos that "perpetuate the myth that desirable human traits are transmitted genetically, not socially, and that the traits most characteristic of certain races and social classes are the most desirable universal human traits."

California Cryobank is sperm world. It feels like a made-for-TV movie, and you can almost hear the background music: "It's Raining Men. Hallelujah!" Men who make the cut come twice weekly to masturbate. Their sperm—thousands and thousands of ampoules of the stuff—is frozen and packaged for shipment around the world.

"The greatest hindrance to medical advances . . . are ethical and legal issues, not scientific obstacles."

Dr. Cappy Rothman is the sperm banker who runs the place. His home, in the hills of Los Angeles, is reminiscent of Woody Allen's in *Sleeper*—ultrawhite and ultramodern, except without the orgasmatron in the corner. Instead he has a lot of penis sculptures, a few real penis bones (from mammals that have penis bones), and an impressive collection of penis-related gag gifts. What appear to be Beanie Babies of whales lining a bookshelf are cuddly sperm toys, a gift from Fairfax Cryobank, a competitor in Virginia. He also has pens that have wiggly, light-up plastic sperm; sperm soap; and "Dick-Tacs," penis-shaped candies in a container resembling a Tic Tac box. And that's only a few of his jewels.

Rothman has a gray goatee, a slight paunch, and thick, black glasses (I wasn't sure if they were old-fashioned or high-fashion retro). He is nearing 70 but looks at least ten years younger. He is affable and likes to gab. Though he is from the Bronx, his slow, easy-going pace is definitely more LA than New York City. He talks as if he is playing a word association game, flipping from subject to subject without any apparent reason for the transition. He talked about the early years in the sperm banking industry, then showed me photos of his new condo in New York City, then talked about his Japanese rock garden with a waterfall that he plans to expand, then talked about his one-year medical fellowship studying erections, then told me that he donated three vials of sperm to the gay and lesbian gala benefit (not his own sperm; the winner gets to choose any three vials from California Cryobank, worth about $1,000).

He has a copy of an artificial insemination documentary in French on his computer that he likes to watch in fast-forward. It's all talking heads with no subtitles, so unless you speak French, you have no idea what's going on. But Rothman slows the film to normal speed when he is on camera speaking in English, so you can watch Dr. Cappy Rothman watching Dr. Cappy Rothman saying the same things about the sperm business—word for word, in fact—that he tells every visitor in real life.

Next to the bookshelf decorated with his sperm toys, he has photos of himself and family members with celebrities. There were also lots of shots of his family holidays around the world and a huge photo collage that he made for his 40th anniversary. He has three handsome sons who would definitely make the cut at the sperm bank—based on looks—though he will not let his sons donate there. With admission standards so high, it would seem like nepotism if they got in, he said, though he would have no problem with them donating elsewhere. If he had daughters, he said he would "be happy for any of them to hook up [with a donor]. We get to know these guys pretty well. They are nice young men." There was a photo of two of his sons with their arms around the Dalai Lama. There was a photo of Rothman with Phil Donohue. He was on *Oprah*, but there was no photo. Rothman has been in the frozen sperm trade since the 1970s and has a history of doing really weird things with sperm, which got him into the tabloids and on talk shows. He made headlines for retrieving sperm from dead men. (In case you're curious, you have to rush to the morgue and squeeze out sperm from the epididymis or testicles. Sperm survives about 48 hours after death.) He was not the first to do this, but he is, as one writer called him, the ringleader. One Rothman client gave birth to a daughter with her husband's sperm four years after he died. Rothman had retrieved the sperm immediately, and they kept it frozen until she wanted to use it. The notion rankles many. But Rothman wrote in a medical journal that "it could actually be unethical to deny the hope and help available through sperm retrieval to a grief-stricken wife on the basis of the physician's subjective morality of right and wrong (good and evil)." Lori Andrews, a lawyer who specializes in reproductive rights and a critic of donor sperm from the dead has called Rothman and others of his ilk sperminators.*

The white-capped ampoules are for Caucasian sperm; yellow caps for Asians; black caps for African-Americans; and red caps for mixed breeds (mutts).

Rothman has no patience for the ethicists who grapple with the morality of it all. "The greatest hindrance to medical advances," he once wrote, "are ethical and legal issues, not scientific obstacles." He calls the FDA the great gorilla. He wants to expand his sperm bank into China and Russia.

From the outside, California Cryobank looks like a small suburban strip mall—a modest building of grayish brown gravelly stones with large forest-green awnings. It looks like it could easily house a Starbucks and a stationery store.

---

* In *The Clone Age*, Lori Andrews writes about all of the bizarre experiments, including using electroejaculation, which she says is like a cattle prod, to force sperm out of comatose men.

Sperm samples at California Cryobank, Los Angeles, California. White caps for Caucasian men, yellow caps for Asian men. Randi Hutter Epstein, M.D.

From the inside, California Cryobank resembles a UPS storage facility. The basement is a large room filled with packages wrapped in brown paper for delivery. There are vats of frozen sperm in one corner, frozen embryos in another, and on the other side of the glass partition, stored umbilical cord blood. Many banks have expanded into multiuse facilities. (Once you have the freezing capabilities, you can freeze anything and make more money.) The walls are decorated with huge colorful photographs of pregnant women and sperm.

One of the sperm packagers was a large, friendly man from Eastern Europe. He lifted the cover of a sperm vat containing ampoules of sperm. Fumes of cold liquid nitrogen wafted upward, transforming the room from mundane to creepy. (Dr. Rothman insisted he get a photo of the two of us together smiling near the vat. He has a bulletin board filled with photos of himself and every journalist who has ever visited.) Each vat, about the size and appearance of a beer keg, contains 20,000 ampoules of sperm. Each ampoule, about half the size of a cocktail straw, contains about 60 million sperm. You don't need to do the math to figure that when you are in the basement of California Cryobank, you are in a winter wonderland of sperm. The white-capped ampoules are for Caucasian sperm; yellow caps for Asians; black caps for African-Americans; and red caps for mixed breeds (mutts). The whites were the majority. Sperm bankers say they are the most eager donors. No one knows why. A writer once accused the banks of not soliciting minorities, but that's not the case at all. They yearn for variety. What it means is that if you are applying, your chances of getting accepted are much better if you are ethnic. Rothman says they usually nix anyone shorter than 5 feet 9 inches, but he will take a short full-blooded Italian. Italian sperm are highly desired.

California Cryobank has three branches. Rothman lives near the UCLA branch, the main one, but they also have a recruitment center near Stanford University, in northern California, and one between Harvard and MIT, in Cambridge, Massachusetts. It is no mistake that they chose locations near top universities, looking for intellectually advantaged sperm. But Rothman says he is toying with the idea of sperm from the noble professions. Noble, not Nobel. That was already tried. From 1980 to 1999, a wacky and wealthy dentist opened a sperm bank touted to store sperm from geniuses. In the page-turning book *The Genius Factory: The Curious History of the Nobel Prize Sperm Bank*, David Plotz traces the average and less-than-average kids born from brainy donors. Rothman (who has a photo of himself with Plotz on his bulletin board) told me he was offered the chance to adopt the genius collection after the owner died. He didn't. Rothman, for his *noble* collection, is

Masturbatorium at California Cryobank, Los Angeles, California. Randi Hutter
Epstein, M.D.

thinking firemen, policemen, and paramedics. He wants to assess
demand.

Less than 1 percent of men who apply to California Cryobank
are accepted. As journalists love to report, it is easier to get into
Harvard than a sperm bank. Actually, it is easier to get into a sperm
bank if you are already in Harvard, but not vice versa. You have
to be smart, handsome, pass muster on several interviews that
delve into family history, and eager to masturbate twice weekly
for about a year. Other banks make similar claims about the difficult application process. Rest assured, someone is going to launch
a business that caters to prospective applicants—downplay the
alcoholic uncle but talk about the cousin who played football for
Michigan.

The main floor, the one above the storage facility, is for inter-
viewing and masturbating. The masturbatoriums (their name)

look like small, sterile powder rooms minus the toilet. They are
decorated with posters of naked women. There is one small TV
to play videos and a stack of pornographic magazines. Rothman is
particularly proud of the room he decorated. One bored, pimply
faced young man was sitting in the waiting room.

Administration and marketing are on the next floor. Young
attractive women interview candidates and reject the ones they
decide are ugly or will not be reliable masturbaters. (How do they
judge that?) The women, who are called
donor coordinators, say they judge men first
by sperm quality, second by the answers on
their questionnaires, and lastly by personal
interview. Each candidate is given a chart
that looks like a seven-step food pyramid,
with a large base and tiny top to emphasize
how difficult the process is. Step one, the
base, begins with the basics (height, weight,
age, sperm quality). You must be between
19 and 39 years old. The top step is the final
review and approval by the medical director. On the bottom of
the page, it says: 1,000 donor applicants. On the top of the page,
it says: 9 donors complete the process. Rejects are sent an e-mail
without stating the reasons for the cut. They do not want candidates to try to improve their performance and reapply. There is
no waiting list.

> "In the old days people were happy for a solution to infertility." Now lesbian couples "want the ideal man."

Administrators claim that after several interviews, including
one with a genetics counselor, they glean honest information.
I have my doubts. Most applicants are undergraduates, and the
$900-a-month prize for masturbating seems like enough of an
incentive to forget about a family history of a potentially heritable
disorder.

Most married couples choose sperm whose donor looks like

the husband to keep the process a secret from outsiders and perhaps from the child. Single women and lesbians can choose any kind of man they desire. How does a woman decide? For those who hate going through the racks, California Cryobank provides, for another fee, a personal shopper. La'Trice A. Allen is the consultative services manager. For $60, she will have a half-hour telephone conversation with you after you have scanned the catalog and narrowed your choices to about six men. For $120, you can select a dozen men and she will meet with you for an hour to go over them. For $300, she will have a telephone conversation with you and go through the catalog for you. And for $500, she will meet with you in person and go through the catalog, truly doing the shopping for you. Allen sees the photos of the men matching the donor number, but cannot show them to the client. Some women will bring magazine photos of men they find attractive so Allen can get an idea of their likes and dislikes. She's been doing this for 17 years. "In the old days," she says, "people were happy for a solution to infertility." Now, with single women and lesbian couples, the buyers are "more choosey. They're demanding and aggressive about what they want." Lesbian couples, she says, "want the ideal man."

Rothman says they choose according to market demand. The best catch, nowadays, according to the women they interview, is a 6-foot college-educated man with blond or brown hair, blue or green eyes, and dimples. Dimples are important.

Sperm is expensive. An ampoule costs $340 for intracervical insemination and $400 for intrauterine insemination, which requires further processing. Most women buy a few vials at a time because there is no guarantee the first shot will work. You can also buy vials to save for future siblings from the same donor.

Like all sperm banks, California Cryobank likes to hear about pregnancies, the success stories. Rothman encourages buyers to report pregnancies for their own good; should a genetic defect arise in a half-sibling, they would be able to inform other parents. Alas, reporting is voluntary and few women make the call.

"We consider it more of a tissue donor, like getting a blood transfusion, not a father."

Perhaps after going through so many specialists—the bank, the fertility doctor, the obstetrician—and then the chaos of a new baby, parents forget.

Catherine and her female partner each wanted to get pregnant. They went in age order to favor the biological clock. Raised in an Orthodox Jewish family, Catherine chose a fertility specialist known for working with Jewish clientele. He is *Fiddler on the Roof*'s Yente picking the suitable specimen. Catherine thought the experience would be female bonding, waiting in a doctor's office with a group of nervous women poised to begin the same process, much the way her friends bonded with other parents during an overseas adoption trip. Not so. She was the lone lesbian among a group of humiliated Hasidic women.

And then the next shocker. The doctor told Catherine that he did not deal in Jewish sperm. Orthodox Jews have to use sperm from non-Jews. Jewish law permits artificial insemination (Catholics and Muslims still have problems with it). Yet according to the Orthodox rabbis, sperm must be from a Gentile donor to avoid the rare chance of marrying a half-sibling. The rabbis figured that there is no chance an offspring would marry a non-Jew, so that eliminates the possibility of incest. Any child born from a Jewish mother is considered Jewish, so there is no question about the religion of a baby born from Gentile sperm and a Jewish egg.

Catherine shopped elsewhere because she wanted Jewish sperm. "I couldn't go back to my baby's grandmother—my mother—with a blond-haired blue-eyed shiksa kid." She got her Jewish sperm. "And then what happens? I give birth to my blond-haired, blue-eyed daughter."

Catherine now has two children (she got pregnant and so did her partner), both from donor 1355 (number changed), from the Midwest. They chose him because he said in his essay, "I am Jewish by birth and by choice." He also said he was good with his hands and loves to sail. Their daughter—his sperm recipient—likes to sail too. "We had a choice of an Israeli who was 5 feet 2 inches with great motility and a good statement. But we went with 5 feet 11." She went to a bank whose slogan is "science with a human touch," which I thought a rather crude pun considering all the masturbation.

The process, for Catherine, was harrowing. It took 11 cycles, using four vials per cycle, plus lots of fertility drugs to boost the odds of conception. "You go to the hospital when you're ovulating and you have this password and secret number. They ultrasound you and then you go and with your ID, you pick up the sample. It was a 2-inch plastic tube. It's nothing. I walked with this tube under my armpit or else they'll die. They say armpit is the proper temperature for sperm. You sign in and you have your sample and you hold it. It's like signing a baby out of a hospital. Then you're in this waiting room." Catherine was told her donor can only provide for six live births, and she has two of them. But you never know.

They paid top dollar to get all the information they could but have no desire to know him personally. Unlike so many thousands of children out there searching for their biological fathers, Catherine said neither of her children ever expressed any desire. "We consider it more of a tissue donor, like getting a blood transfusion, not a father. We didn't choose the California Sperm Bank because back then you had the option of opened or closed [donors] and I wanted nothing to do with that." Some banks offer men the option to donate and allow the prospective parents or offspring to contact them. Few men take this opportunity, which is understandable because you are turning your donation into a family situation. As Catherine says, "This was just a donor. We did not want a third wheel. She [their daughter] has never thought about it by anything other than a donation. We don't treat it like a human being. It's just another sample."

Catherine believes that the media buzz over children hunting for their donor dads—there are several Web sites that facilitate connections—exaggerates the numbers of children who truly want to connect. It's hard to say. Half-siblings who connect make for great daytime TV, and a few have gotten on *Oprah*. Are they the norm?

Wendy Kramer could not disagree with Catherine more. She and her son, Ryan (made from donor sperm) founded Donor siblingregistry.com, a Web site facilitating matches among donor offspring. She says she has more than 15,000 members and has had thousands of matches since she began in 2000. In Februrary 2007, her son became the 2,910th person to find his half-sibling. Wendy e-mailed me a photo of Ryan and his half-sister. It was a wonderful family reunion of sorts. Wendy was so touched when Ryan's half-sister sent her a card on Mother's Day. "Donated tissue?" she says. "To my son, this donated tissue is half of his genetic identity."*

> Even Aldous Huxley's fictional *Brave New World* babies needed finishing touches after birth to ensure they would grow up according to their pre-ordained biological destiny.

---

* According to an NPR report, in August 2000, the California supreme court decided that anonymous donors do not have unlimited rights to privacy. The decision was based on the court case regarding the child born from donor sperm that had a family history of a serious kidney disorder. He had donated more than 300 times before being rejected for his medical history. The issue is, when do you have to come forth? Will the court's decision dissuade men from donating? Already

The whole notion of buying a set of genes cuts to the core of the nature-nurture debates. How much of your identity do you believe is shaped by genetics? Are your genes the blueprint for the person you are to become or a rough draft constantly rewritten and edited every day? Even Aldous Huxley's fictional *Brave New World* babies needed finishing touches after birth to ensure they would grow up according to their preordained biological destiny.

Mechanical fertilization, as Sims called it when he wrote about the real stuff going on, has arrived. Sperm shopping is no longer a clandestine service. To be sure, we've come a long way since the days of a doctor secretly inserting student sperm into an unsuspecting wife, and yet it seems we still have a ways to go to figure it all out. Will we implement regulations, and what sorts? One thing is certain. When it comes to the sperm trade, the mechanics have been worked out in easy-to-follow steps. Masturbate. Freeze. Store. Ship. Thaw. Insert.

---

stringent policies in Europe about open donation have reduced supplies tremendously. Stephen Bayer, an IVF specialist from Boston, told NPR's Tovia Smith that he worried that donors will refuse to participate. He said there is no such thing as the perfect donor. Everyone has a family history.

14

The Big Chill

"If we get a good embryo, are you planning to be the carrier?" The doctor was talking to a 38-year-old slightly sedated woman who seemed perplexed by the question. That was why she was there in the first place, because she desperately wanted to be pregnant. The doctor was thinking about surrogacy, lending the eggs to another woman to gestate. When the patient said that yes, she planned to be the carrier, the doctor did a so-called mock insemination to make sure the paths were clear.

Five women—a doctor, a fellow, a nurse, a technician, and a patient—were together in a high-tech room with a surprisingly old-fashioned feel. This was a modern version of colonial birthing rooms, and the "birth attendants," or rather helpers, played the twenty-first-century role of the gossips (or God's sibs), the name for female companions comforting the laboring woman. Just like our great-great-grandmothers must have done, we chatted about mundane things—career and kids—all the while trying to ease the fears of the expectant woman. In this case, though, expectant meant that the woman expected to become pregnant someday, not that she already was. And in this case, they were in a sterile room at Yale University, wearing surgical garb and masks and observing the doctor as she used ultrasound to guide a skinny straw inside the birth canal and suck fluid from each follicle, hoping to snatch an egg. The fluid-filled tube was passed immediately to two embryologists through what looked like a take-out window, (a slot that separated the procedure room from the laboratory).

> "I said I froze my egg and the doctors said to me, 'you made a mistake, you mean you froze your embryos'"

As they got the sample, they examined it under the microscope and would call out the score—"one good one"—and so on, until they got a total of 14 eggs, considered a good harvest.

When they finished, after about 30 minutes, the embryologists examined the eggs with a microscope—a cloud formation of grayish blobs and specks. It was all so shockingly nonchalant, so understated, so muted. The embryologists siphoned away the bubbles surrounding the egg with what looked like a tiny turkey baster. They needed each egg solo to assess maturity—how many divisions it had accomplished.

Half the eggs were mixed with donor sperm. The rest were frozen. Freezing eggs is problematic because they do not thaw well. Embryos do better in the freezer, but not every woman has a batch of sperm available, and some do not want to use donor sperm.

In the beginning—less than ten years ago—egg freezing was a rare and experimental procedure for women with cancer. In 2001, for instance, Lindsay Nohr Beck was told that the tongue cancer she thought she had beat just two years earlier had returned and spread to her lymph nodes. Doctors said she would survive, but chemotherapy would likely render her sterile. "I know this sounds

kind of stupid, but I first heard about egg freezing from the movies. There is this one scene in *You've Got Mail* when she says she is
going to have her eggs harvested. I knew they froze sperm." Beck
found willing investigators at Stanford University who froze 29
of her eggs.*

The catch—and it's a big one—is that no one knows whether
the process works until the eggs are defrosted. Sometimes freezing destroys eggs, but you can't tell the bad eggs from the good
ones in their frozen state. After Beck made her deposit, fertility
experts said, for the record, that they hoped medical technology
would advance by the time she needed to thaw her eggs. Off the
record, they said the girl was banking on a pipe dream. It's hard to
tell a woman diagnosed with tongue cancer twice not to count on
modern technology.†

Beck knew the doctors were pessimistic. "At my first American
Society of Reproductive Medicine meeting I said I froze my egg
and the doctors said to me, 'you made a mistake, you mean you
froze your embryos.' A lot of them told me it would never work.
I went home crying."

Today, egg freezing has gone the way of all new reproductive
techniques, from a bizarre idea to a commercial enterprise. Everybody is in the business—there are labs across the United States,
Italy, South Korea, Australia, Japan, and Israel. The skeptics five

* Few investigators were working with oocyte cryopreservation, though the
idea was intriguing because it enabled doctors to work with gametes, rather than
embryos, and avoid the ethical implications. One of the first successful thaws of
an oocyte resulting in a pregnancy was by an Australian researcher who reported a
twin pregnancy in 1986. C. Chen, "Pregnancy after Human Oocyte Cryopreservation," *Lancet* 1, no. 8486 (April 19, 1986): 884-886.

† According to American Cancer Society's *Cancer Facts and Figures 2007*, 2,870
women were diagnosed with tongue cancer and 650 died from it. Compare this
rare cancer with breast cancer that was diagnosed in 178,480 women in 2007 and
killed 40,460 during the same year.

years ago are now among some of the industry's greatest devotees. Thawing has improved, but it's far from perfect. Like sperm, no one is keeping track of the numbers, and so statistics are murky. Experts say upward of 1,000 babies have been born from frozen eggs. Though it is still considered experimental and success rates are depressingly low, triumphant stories that hit the press every so often are beginning to lure women, each paying upward of thousands of dollars, each gambling on an imperfect science.

Curiously, the vast majority of customers are not cancer patients, who originally inspired the research, but healthy single women, like the woman at Yale, who for one reason or another want to put their eggs on ice in a Dorian Grey state as the rest of their bodies confront the inevitable. Haven't found Mr. Perfect yet? Want to focus on a career without waking up in the middle of the night to nurse a newborn? Why not ice an egg and reheat it when the time is just right? The way the marketers tell it, you are better off freezing your own egg than buying a fresh one from a younger woman later on. Of course, regular intercourse may do the trick too, but as fertility experts remind us, you cannot rely on your biological clock. Who knows, maybe one day we will be able to reset the clock and let it run as long as we need.

Let's be clear on one thing. Egg freezing is not about masturbatoriums filled with photos of sexy men and porn videos. Whether you are freezing eggs to donate to another woman (and getting paid a lot of money) or freezing eggs for your own future use (and paying a lot of money), it is not fun. We're talking about injections of potent hormone drugs that hyperstimulate your ovaries to spew dozens of eggs in a cycle followed by an invasive procedure to retrieve them. While sperm banks are turning eager masturbators away, it takes marketing prowess to encourage women to go this route.

The egg is "the most amazing cell in the body. They are rare, hard to get to and they are fragile and extraordinarily robust at the same time," explains Dr. Teresa K. Woodruff, chief of the Division of Fertility Preservation at the Feinberg School of Medicine, Northwestern University, in Chicago, Illinois. "It's a cell of extraordinary contradictions." Woodruff, a renowned basic scientist, is also a professor in the department of biochemistry, molecular biology and cell biology and in the department of obstetrics and the founder of the Oncofertility Consortium.*

"They are always in an 'oh my God, I'm ready, I'm ready' state."

Women start making eggs at 9 weeks of gestation. Not 9 weeks old. Nine weeks after conception. It is a stunning feat. By 5 months of gestation, we have made upward of 7 million oocytes, the immature eggs. The vast majority are dead by birth. They committed suicide (doctors call it apoptosis). No one knows why. We are left with a million that continue to die after birth. From their grand beginnings, a virtual empire's worth of oocytes, a woman ovulates a mere 400 to 500 in her lifetime. I hate to keep comparing us to men, but our entire supply is less than the average number of sperm per ejaculate. To my way of thinking, that makes each egg a precious commodity. One human egg sells for thousands of dollars more than a sperm.

A human egg is 100 micrometers in diameter, the width of a strand of hair. That doesn't sound like much, but it is the biggest cell in the body, filled with fluid and delicate paraphernalia, ten times bigger than a sperm, which is basically a little head with DNA. Every egg is housed within a follicle that is lined with electrically charged cells, ready to release the egg into the ovary's

---

* Woodruff has published extensively on the basic science of the egg and the blossoming field of "oncofertility," which means trying to consider the fertility issues of women while they are undergoing cancer treatment. See the bibliography for a smattering of some of her papers.

version of the outside world. Unlike any other cell in the body, oocytes remain in a sort of oocyte adolescence until the moment of ovulation.

Dormant oocytes are particularly vulnerable to damage. For one thing, they do not have repair mechanisms that other cells have. Moreover, they reside in a highly excitable state. "When you are that ready, even the slightest push in the wrong direction or a slight temperature change, or a little electrical shock can release a cascade of events," explains Jason Barritt, an embryologist and scientific director of Reproductive Medicine Associates in New York City. (Few men, I must add, get as excited about oocytes as Barritt.) That may be why older women have more trouble conceiving—too many of their eggs are damaged goods. On average, the percentage of healthy eggs drops from 90 percent among 21-year-olds to 10 percent among 41-year-olds.

As Barritt puts it, "They are always in an 'oh my God, I'm ready, I'm ready' state. The moment the channel opens, the cells are like 'Woo Hoo! We are on our way!'" Typically, only one escapes the ovary, a process called ovulation. Nowadays, fertility drugs that put the system into overdrive send dozens of eggs shooting out of the ovary.

Immediately before escaping the ovary and bursting forth into the outside world (relatively speaking), they do their final dance into maturity. Two strands of DNA separate so that each cell has only one strip of 23 chromosomes ready to meet its mate.* (During the final split from a 46-chromosome egg into two 23-chromosome

---

* Researchers have been trying to coax immature eggs into maturity—the hope being that doctors could retrieve the immature eggs from a woman and use those to make a baby. Woodruff's team at Northwestern University, for instance, reported on October 6, 2007, in the *Journal of Tissue Engineering*, that they were able to take immature eggs from prepubescent mice and grow them inside a follicle with a special medium that allowed the cells to mature, and one egg spawned a pregnancy.

eggs, only one mature egg gets enough cytoplasm to mate. The other one is a clump of DNA that disintegrates.)

Woodruff is one of a handful of scientists trying to decipher the follicular signals telling a particular egg to finish maturing and ovulate. "You can have two follicles right next to each other," she says, "and one gets a signal to ovulate when the woman is 19 and the other waits 20 years."

Concocting the perfect ingredients to freeze eggs without forming ice crystals has become the Holy Grail of the fertility industry. A single strand of 23 chromosomes dangles in the center of the eggs, held in its precise position by a network of threadlike spindles. One ice crystal poking at the innards can decimate the supporting structure and send the DNA into a tizzy. Several teams in the United States are in a tight race to patent formulas. If they ever come on the market, not only will women have to decide whether to freeze their eggs but they will have to weed through the marketing campaigns to decide which freezing method to choose.

Though details vary, there are basically two ways to freeze an egg: slowly or quickly, not in between. Either extreme minimizes ice crystals. The slow freeze takes about an hour and a half. The temperature drops gradually, allowing ice crystals to form outside the cell, not inside.

The newer rapid freeze is called vitrification because it forms a glass-like casing around the egg. Technically, the egg is not frozen because there is no ice. It is vitrified (converted into glass) and rewarmed, not frozen and thawed. So far, the vast majority of babies born from frozen eggs come from the slow-freeze technique, but increasingly, doctors are turning to vitrification. One embryologist told me that vitrification has to be done with a sense of urgency, so quickly you don't even have time to answer the telephone, which makes me wonder why an embryologist dealing with a woman's delicate egg and her potential future baby should even consider answering the phone in the midst of it all.

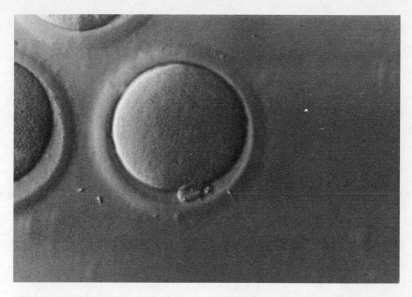

A sperm entering an egg. Courtesy of Vanessa Guelman and Pasquale Patrizio, of Clinica Genese, Salvador, Bahia, Brazil.

There are similarities between the two recipes: Both methods dehydrate the cell and replace it with a sugar/alcohol or salt/alcohol cocktail. The egg marinates in a highly concentrated sauce that lures water out of the cell as the alcohol seeps in. The cocktail is called a cryoprotectant, as in protection from freezing. The alcohol is not the same kind used to mix cocktails, but it is toxic. Eggs remain intoxicated until you are ready to use them. Then, every drop of cryoprotectant must be removed from the egg before it mates.

Nothing about the egg freezing/thawing business is predictable. You cannot even tell if the egg has been damaged until it defrosts. So you can be paying for storage for dead eggs and not know it.

Both methods create a thawed egg with an excessively hard shell, so hard that the sperm cannot break inside. The frozen-egg industry was helped tremendously by another fertility technique developed to help men with sluggish sperm. In the late 1980s, fertility specialists started poking holes in eggs to nudge slow sperm inside. The method is called ICSI (pronounced "icksee") and stands for in intracytoplasmic sperm injection. ICSI is the only way to get sperm inside a thawed egg.

In the past 30 years, the number of women in their early 40s having their first child has catapulted 25-fold. The escalating rate reflects, in part, an increasing number of women postponing motherhood because of careers or because they have not found the right partner. It also reflects increasing success of fertility treatments. For egg-freezing entrepreneurs, the growth of the older-mother brigade means a growing client base—customers who may be interested in buying someone else's frozen eggs or paying to freeze their own. The early impetus, to help cancer patients like Lindsay Nohr Beck, has been eclipsed by bankers eying a lucrative market among single women.

"The pill allowed women to decide when they did not want to become pregnant. This is allowing women to decide when they want to become pregnant."

Dr. Jeffrey Boldt, an embryologist at Community Hospital in Indianapolis, is a medical director of Cryo Eggs International, touted as the first for-profit frozen-egg donor bank. They ship eggs worldwide. Eggs travel like sperm, each tiny packet cushioned in a vat that is about 2 feet high and 1 foot wide. The inside is kept at a precise cold temperature so the eggs do not rot en route. "They travel alone," says Diana Thomas, the president and CEO, "just like sperm and embryos." At the other end of the journey, they are met by a customs broker who chauffeurs them through paperwork.

Boldt says that buying frozen is better than buying fresh. Frozen is cheaper than fresh, costing about $15,000 per cycle compared with $25,000 to $40,000 for fresh. Better yet, when you buy frozen, you and the donor do not have to synchronize menstrual cycles, a hassle during typical egg deals. When you buy fresh eggs, you pay in advance and hope the donor will produce a plentiful batch. On the downside—a huge downside—you are less likely to get pregnant with a batch of frozen eggs. About 3 to 5 percent of frozen eggs yield babies, compared with 6 to 8 percent of fresh eggs.*

So far, nine babies have been born with eggs purchased at Cryo Eggs International. The bank boasts a 70 percent egg survival rate, of which 69 percent fertilize. After that, you are dealing with the same odds as IVF. That's about a 30 percent pregnancy rate per cycle. Cryo Eggs suggests that customers increase their chances of a take-home baby by purchasing about a half dozen eggs, which should make two to three embryos.

The Yale team has had 34 women freeze eggs so far, about a quarter of them because of illness. The cost of harvesting eggs is $6,500. They do not charge for storage.

The Donor Egg Bank, based in Los Angeles, has a guitar instrumental introducing its Web site and their logo is "where dreams meet innovation." They deal in fresh and frozen donor eggs. You buy a frozen batch (eight eggs) for $12,750. An additional $500 allows you a second batch for free if the first one is a dud. Just like the sperm trade, women can flip through catalogs of potential donors, though most of the donors have checked the box that says

---

* Eleonora Porcu's lab at the University of Bologna, in Bologna, Italy, reported the first baby born from a frozen egg that used the ICSI technique in 1997. E. Porcu et al., "Birth of a Healthy Female after Intracytoplasmic Sperm Injection of Cryopreserved Human Oocytes," *Fertility and Sterility* 68, no. 4 (October 1997): 725–726.

willing to donate fresh eggs but do not agree to donate eggs for the freezer.* "Originally we only wanted to deal in frozen," says Brigid Dowd, the director. "But we found that because it's a new technology, a lot of people didn't want to do it, or their doctors didn't want them to do it. You know, it's new, scary, they want to do it the old-fashioned way." (In this case, "old-fashioned" referred to getting pregnant with another woman's fresh eggs.) She says that they invested a lot of money in frozen-egg techniques and then had only ten donors and no one wanted them. "In the last six to nine months, everyone has started to want frozen. We kept the prices low to give the patients a break in exchange for trying a new technology." The Donor Egg Bank offers a "concierge service" that provides assistance with flights, transportation, and hotels.

Rather than selling other women's eggs, Extend Fertility, based in Woburn, Massachusetts, freezes women's eggs for their own use later on. As Allison Aubrey put it on National Public Radio, Extend Fertility founder Christy Jones "saw opportunity in the ticking biological clocks of her 30-something Harvard MBA classmates." Jones likes to say that she was her own first customer, and yet shortly after the freeze, she met a man, married, and got pregnant naturally. Jones is not a scientist but lured leading fertility specialists throughout America to partner with her. Extend

---

* Better than the sperm banks, the Donor Egg Bank provides photos of each donor and explicit information online. I was not shocked that the bank would collect this information but that women would allow their photo, age, weight, and minutia of their and their family's medical history on the Web for anyone to see. It gave me the creeps. Donor 850991, for instance, is a pretty 5-foot 8-inch blonde, who scored (according to herself) 1,500 on her SATs, and includes cheerleading under "academic information." She also reports that her grandfather has Alzheimer's, her sister suffers from seasonal allergies, and she can type 95 words per minute without mistakes. She also says that her favorite book is *Moby-Dick* and her unique learning experience was figuring out that "being popular in high school was not the greatest thing in the world."

boasts a postthaw survival rate of 85 percent and a 61 percent pregnancy rate. The numbers are small. So far, they have had 11 deliveries, including three sets of twins, and one ongoing pregnancy. They say they have more than 250 clients (excluding the cancer and medical patients), each investing $10,000 to $13,000 for egg retrieval, $400 for the first year of storage, and $75 annually thereafter. An Extend spokesperson told potential clients at a March 2009 meeting in Manhattan that they accept credit cards, so you can get a lot of miles.* Everyone signs a consent form stating what should become of the eggs should they die first or if they decide not to use them.†

Not everyone is gung-ho. Many doctors, including a few potential competitors, say it's too soon to turn frozen eggs into a

---

* They were apparently talking about airplane miles that you can accumulate using your credit card, not speaking metaphorically about extending the "miles" on the life of your eggs. Someone in the audience asked a client who had just given a testimonial about Extend whether one should admit to dates that they have frozen their eggs. I was mystified by the discussion this question provoked, which led to talk about whether to put "egg freezer" on your online dating profile and how many dates you should wait to bring up the topic. I mean, what man wants to hear about a woman's gynecology appointments on the first date, or second or third for that matter? And why do you need to get into any kind of gynecological history on a dating Web site? As for the Extend prices, they vary according to center. Austin is one of their cheapest, charging $720 for the initial consultation, $10,803 for retrieval, $3,000 to $5,000 for drugs. New York is at the high end, charging $525 for the initial consultation, $13,055 for the retrieval, and again, $3,000 to $5,000 for the drugs. Women with cancer get a discount if they work with Fertile Hope, their nonprofit partner. Marla Librati, vice president of marketing, said that customers should realize when they are shopping around that other clinics often try to "low ball their prices but don't include anesthesia and then charge high storage fees."

† A famous case, described in Debora Spar's *The Baby Business*, involved a millionaire couple who stored embryos in Australia and then died in a plane crash. The clinic heard from several eager women willing to "adopt" the embryos and hoping for the rightful inheritance.

moneymaking enterprise. Dr. Pasquale Patrizio, director of Yale REI and Fertility Center, says, "If you asked me a few years ago, I would be disappointed to have companies making money since the procedure was still highly experimental and few births were reported. Today the technologies are much more advanced, and the numbers of births from frozen/thawed oocytes have been increasing. I am, however, still unsure of what is the right price tag. Some companies are selling frozen eggs with a high price tag per egg, as much as $2,000 an egg. People should realize that not every egg you freeze makes a baby. It's more like one in ten, and this scientific fact should be taken into consideration."

Critics worry about women being sold a raw deal. The American Society of Reproductive Medicine considers egg freezing experimental and in 2007, much to the dismay of the frozen-egg entrepreneurs, said egg freezing "should not be offered or marketed as a means to defer reproductive aging" because not enough evidence confirms its safety and efficacy. Charisee Lamar, director of the National Institute of Child Health and Human Development's Neuroendocrinology Program, agrees: "For adult men, sperm banking is tried and true. If you are an adult female and you are with someone who can be a sperm donor, embryo cryopreservation is tried and true. Everything else for adult women is experimental, including freezing whole ovaries, pieces of ovaries, and eggs. There are options that are becoming available but women need to be cautious." Skeptics worry about long-term effects on babies born from frozen eggs. There have not been any large-scale trials.

Reproductive entrepreneurs scoff at the naysayers. They compare egg freezing to the early days of IVF, when doctors and patients worried about the medical, ethical, and moral ramifications. As Alan B. Copperman, director of the Division of Reproductive Endocrinology at Mt. Sinai Medical Center in New York City and an Extend Fertility partner, says, "It has the potential to be the most liberating advancement for women since the birth control pill. The pill allowed women to decide when they did not want to become pregnant. This is allowing women to decide when they want to become pregnant." Except for one thing: success rates are still depressingly low.

The issue, Debora L. Spar explains in *The Baby Business*, is not whether to sell frozen eggs but whether to regulate the sales. Egg freezing is a new science, she adds, that screams for regulation but bumps up against a medical community and public that hate being regulated. "Unlike its European counterparts, the U.S. government is notoriously loath to intervene in markets spun off from high-technology sectors or to impose regulatory constraints on high-growth industries," she writes. Spar likens the baby market to the early days of radio and telegraph and the Internet that "moved from an initial period of market anarchy to an eventual demand for rules." She believes that the children of donor eggs and donor sperm will push for regulations because they will want to know their genetic history.

In January 2008, a group of embryologists, reproductive endocrinologists, and frozen-egg bankers met in Atlanta to discuss launching a frozen-egg registry to track the children. Lindsay Nohr Beck attended the meeting. She came not as a client but as a representative of the not-for-profit she founded called Fertile Hope that provides fertility information to cancer patients and raises money for research. Yale's Patrizio was there, too. He said they bounced around ideas, including how to monitor patients, who should be included, and what sorts of questionnaires parents would be asked to complete. The simple yet crucial bits of information are: How many eggs frozen? How many thawed? How many babies? How are they doing? At least they are considering a framework of organization that never was discussed in the sperm business.

When the *Wall Street Journal* printed an op-ed piece explaining the frozen-egg business, a reader responded: "Regarding long-term human egg storage, I have a few simple questions: Are you nuts? Can anyone really imagine that a generation of 50- to 60-year-old first-time mothers is a 'solution' to any perceived problem we have put upon ourselves? Has anyone noticed how much energy and work goes into parenthood?"

"What if I meet a guy at 42 . . . and I have no eggs."

In some ways, it seems pessimistic to spend thousands of dollars betting that you won't find the right man in time. Then again, it seems optimistic to have so much faith in science, believing the technique will improve when you are ready for the thaw. Frozen-egg marketers point to the colossal advances in the past decade as a sign that improvement will continue at the same pace. Clients say they are neither critics nor cheerleaders. They are realists who can afford a backup plan should all else fail. They've seen enough friends deal with the sad consequences of trying to procreate with aging eggs. They hope never to use the eggs and will not regret the money spent. They like knowing they have a plan B.

For the most part, women who freeze know the low success rates; and for the most part, their goal is not to start motherhood at menopause. They would like to find someone now. They would like to be happily married. They would like to make babies naturally. Their thinking is that if life does not turn out as planned, they will have a shot using their own eggs rather than using someone else's eggs or adopting. They are, as a whole, optimistic about medical science.

Lucia Vasquez, 33, was eager to spend the money on her eggs.

"I want to travel around the world and I want to do really well in my career. I know it's not a guarantee but it's a really great option. I want to meet the perfect person and then when everything is perfect, that is when I'll have children."

Cara Birrittieri, author of *What Every Woman Should Know about Her Fertility and Her Biological Clock*, encourages all single women younger than 40 to freeze eggs, and if money is an issue, she suggests donating eggs and using the profits to cover personal freezing expenses. As she says, "Talk about a win-win situation."

Copperman, the Mt. Sinai doctor who partners with Extend Fertility, insists, "Nobody is recruiting saying that this is an insurance policy. The message is: we need women to be aware of reproductive aging and incorporate it into decision making. What if I meet a guy at 42 and my FSH is high (a sign of imminent menopause and aging eggs) and I have no eggs. Thank God I froze my eggs at 37. We are not selling pipe dreams. I think women are appropriately counseled. You can say that women should not be sold the experimental technology. But at the same token, if the technology is available, who's to tell a woman she shouldn't."

Given the cost, it is an option for women in a certain income bracket, though Extend Fertility says finance plans are available. "I live in LA. People spend this kind of money on boobs, Botox, and lips without thinking," says Lynn Oddo, an Extend Fertility client. "This is the way of the future. I guarantee that my daughter will freeze her eggs in the 20s. I'm just a woman doing this in the very beginning. I'm a pioneer."

There are other freezing options in the works. A few teams of scientists are trying to freeze immature eggs so that prepubescent girls undergoing chemotherapy, say, can store oocytes that can be nudged into maturing when the women are ready. Other teams are trying to freeze ovaries, or pieces of ovary, that could be transplanted back into a woman and work as good as new.

As for Lindsay Nohr, the two-time cancer survivor, she is pregnant and the mother of a healthy 2-year-old daughter and still has all 29 eggs on ice. Her pregnancy travails, including two miscarriages and detecting a rare defect hindering her husband's fertility, rival any woman's IVF stories. Despite the odds against it, she got pregnant naturally, but says that she and her husband want at least four children, so it's likely she'll dip into her stash at some point.

The fundamental question is not whether freezing will work but what will happen when it does? Will it affect our decisions about motherhood? Will the option to freeze encourage women to delay starting families? And if so, is that a good thing or a bad thing? In Liza Mundy's *Everything Conceivable*, Michael Feinman, a physician with Huntington Reproductive Center in California (part of the Extend Fertility network), says that egg freezing may serve as a cultural pressure (a surprising comment from an egg-freezing entrepreneur). Yet he said to Mundy, "What if law firms someday require female partners to freeze their eggs and wait even longer? Whose interest does egg freezing serve? The woman's or that of an ambitious, still pretty unforgiving culture that doesn't really ever see childbearing for female employees as convenient?"

The egg industry plays into our desire to have everything perfectly planned, beginning with egg meeting sperm. That's all well and good, as we seem to be increasingly able to micromanage conception. And yet, the irony is that all of this obsessive control leads us to one of the most uncontrollable situations ever—parenthood.

Epilogue

There is a spectacular 43-second documentary showing thousands of cells forming the embryo of a zebrafish. It is, perhaps, the closest view of the earliest stages of life ever captured on video. In a color-enhanced version, vibrant splashes of greens, blues, and reds burst into green specks. It looks like a moon. In the final seconds, some dots coalesce in the center into an elongated S. That's the fish embryo. Life is beginning and it's captured on YouTube.

According to the December 2008 issue of *Science* magazine, researchers in Germany made the movie with a new kind of microscope that allows them to magnify and film cells without killing or altering the fish. The investigators played the movie backward to try to figure out which cells turned into what body part. They tracked some 16,000 embryonic cells during their first day of development. There is also a 55-second version with Mozart's Symphony No. 25 in G minor playing in the background.

There are so many minidocumentaries on the Web devoted to baby making. There is one of a sperm being injected into an egg, a fertility procedure called intracytoplasmic sperm injection, or ICSI for short. There are plenty of home videos, too. The BBC posted a few clips from a television show about conception. The show goes back and forth from images inside the birth canal (sperm chasing an egg) to a couple apparently trying to make a baby (two bellies touching). The only scenes of reproductive organs are shot from the inside of them. There is scary music in the background and a narrator, who sounds like a British Don Pardo (the famed radio and television announcer), says things such as: "Inside the vagina, the sperm face the first of many mortal dangers. The walls of a woman's vagina are coated in acid . . . within minutes the walls are littered with corpses of millions of sperm."

Soon enough, there will be no images of conception, pregnancy, and delivery left to the imagination. When a child asks the inevitable, "How are babies born?" there will be no hemming and hawing; just provide them with the Web address. As BettyAnn Holtzmann Kevles points out in *Naked to the Bone*, we have become a visual society. The images feed a curiosity about childbirth and reassure (sometimes falsely) those who have already begun the journey. There is something about keeping an eye on every step of the process that makes parents-to-be feel like good parents already, or at least that everything is going to be okay. They can, if they so desire, choose among a litter of embryos and then monitor the growth ever step of the way and experience the glee of seeing the white flicker on the ultrasound, the first image of the fetal heartbeat. (Of course, there is the other side, when expectant parents are needlessly anxious about insignificant findings.)

And yet, there is another kind of birth image that is popular today and that has nothing to do with diagnosing defects or monitoring the baby, but it has had an enormous impact on the way we perceive childbirth. The recent spate of birth-themed movies, many of them comedies no less, could be considered a cultural response to the current obsession with conception and birth. In 2007, both *Juno* and *Knocked Up* came out, two films focusing on unplanned pregnancy. *Baby Mama*, released in 2008, was probably a first for a comedy about surrogacy. A single career-driven fertility clinic failure hires a trashy, immature surrogate, and the two of them move in together. The odd couple goes twenty-first century—female and lots of baby obsessions. The movie pokes fun at callous doctors, smooth-talking fertility entrepreneurs, and earthy-crunchy birthing-class instructors.

As the Virgina Slims cigarette ads used to say, you've come a long way baby. It wasn't long ago when pregnant Lucille Ball had to fight with CBS executives not only to keep her job but to include a childbirth episode. The word *pregnant* was censored, but they could write *expectant* into the script.

And yet how far have we really come? We're laughing at ourselves—that may be a good thing. But if *Baby Mama* is any indication of our contemporary state of mind, then we still have not figured out how all this fertility stuff should be played out. The movie ends with both women getting pregnant with their own biological babies. There is no surrogacy after all. It all ends happily in an old-fashioned sort of way. Apparently, Hollywood decided, and maybe rightly so, that audiences weren't ready for a surrogate ending.

When the *New York Times Magazine* ran a cover story about surrogacy on November 30, 2008, many readers were enraged. In letters to the editor, they complained about the author, who wrote that she was "practically euphoric" as her surrogate got fatter and less mobile and she went white-water rafting and drank bourbons. The letter writers were also vexed by the cover photo of the fashionable and thin author next to her plain-Jane surrogate. Surrogacy is and probably always will be about rich women paying poor women for their wombs. Apparently, we do not want to read that story or see the images—unless, of course, it can be molded into something funny or with a more palatable ending.

The way men and women decide to have babies and the way they talk about pregnancy has always revealed societal mores as much as it reflects the sophisticated (or not-so-sophisticated) science. Painkillers were developed years before society deemed the drugs okay for use during delivery. Artificial insemination, too, was done years before it was public knowledge and a global enterprise. There will be lots more to come, many things beyond our imagination—or should we say, beyond anything we could ever conceive. Scientists will provide the foundation for the future of baby making (as strange as it may be), but we as a society will erect the framework. And that structure will depend on our current notions of family, women, and sexuality. The laws will follow— there will always be regulations. The ethics will be hammered out—there are plenty of expert panels already. But all of these rules, whether mandated by the legal system or endorsed by a panel of experts, will be constantly reshaped as our society twists and turns through the fertility maze.

Every generation, beginning with the first women who had to suffer Eve's consequences, felt a bit overwhelmed with the advice and the choices. But amid the ever-present confusion, four things are certain.

1. The explosion of choices will make the job of parents-to-be increasingly challenging.
2. Our children and their children and their children's children will have increasingly clear pictures of their growing baby from zygote to newborn.
3. When all is said and done, no matter how much we fine-tune the process, we will still be trying to manage a situation that

is not completely manageable. There will always be surprises, for better and for worse.

4. Pregnancy and childbirth—however you get there—is one of the few adventures you will ever embark on that when you finally get to the finish line, you've only just begun.

Acknowledgments

I like to think of this book the way I think of reproductive technology—I needed a lot of experts—some for their medical know-how, some for their breadth of knowledge on the subject, some for their emotional support, and some for their technological prowess. I feel that this project was a group experience, but of course, I'm responsible solely for the final product and any inaccuracies.

Thanks to all of the librarians who retrieved ancient manuscripts and boxes of scrapbooks and e-mailed me articles. At the New York Academy of Medicine, Arlene Shaner was dedicated from the start, retrieving old books for me and then, at the end of the project, photographing many of the images that appear in the early chapters. (And she kindly e-mailed me the photographs several times because I am technologically challenged.) Winifred S. King, also at the NYAM, tracked down articles about an obscure doctor, Arnold Berthold. I'm not sure how I could have done this without their dedicated document delivery staff, including Stephen Chiaffone, Walter Lipton, Marina Bonilla, Dennis Campbell,

Katherine Molina, and Christopher Borecky, who sent me articles (current and historical) within hours of my request. Chris Warren, PhD, also of the Academy, gave me wonderfully encouraging words of wisdom early in the process. Stephen E. Novak, head of Archives and Special Collections at the Augustus C. Long Health Sciences Library at Columbia University Medical Center, was not only a great help but a great conversationalist, so I'll try to find another project that brings me to his basement haunts. Bob Vietrogoski, who had been Steve's assistant but has since moved on, tipped me off to the Viola Bernard manuscripts, which blossomed into a chapter. I am also indebted to the staff of the Archives and Manuscripts of the Wellcome Library for the History and Understanding of Medicine in London, including Rachel Cross, who facilitated the process of retrieving images; Jack Eckert, the reference librarian at the Center for the History of Medicine at the Francis A. Countway Library of Medicine, an alliance of the Boston Medical Library and the Harvard Medical School; Toby Appel, at the Harvey Cushing/John Hay Whitney Medical Library, Yale University; Sarah Hutcheon, a reference librarian at the Schlesinger Library of the Radcliffe Institute for Advanced Study, Harvard University; Ron Sims, Special Collections librarian, Galter Health Sciences Library, Northwestern University; Jim Gehrlich and Elizabeth Shepard, at the Medical Archives of New York-Presbyterian/Weill Cornell Medical Center. I was fortunate to reconnect with Karen Smith Duffy, a high school classmate (Livingston High School 1980) during my hunt to get a copy of a 1956 article from the *Daytona Beach News Journal*. Karen is the news research editor there.

For their medical expertise, I am appreciative to but do not hold accountable: Jason Barritt, PhD, who explained to me the biology of the human ova with an enthusiasm I never knew a man could have for women's eggs; Joshua A. Copel, MD, who answered thousands of e-mails immediately; Florence Haseltine, MD, PhD; Arthur L. Herbst, MD; Frederick Naftolin, MD, PhD; Sherwin Nuland, MD; Pasquale Patrizio, MD, along with Dr. Patrizio's helpful senior administrative assistant Jeannine Estrada and his gracious embryologist Kathleen Greco; Marcie Richardson, MD, Jeff Riffell, PhD; and Theresa Woodruff, PhD, who is doing some of the most cutting-edge basic science and is a delightfully energetic investigator.

While my children were embarrassed to know that their math teacher came to our home, I am indebted to Jonathan Cuba for sneaking out of school to figure what I was doing wrong with all the digital photographs—why I was receiving high-resolution ones from Arlene and everyone else and sending them out as thumbnails. He's truly a great teacher, had tons of patience with me, and gave me his jump drive when nothing else seemed to store the images accurately.

When I began the book, I knew I needed to learn more about historical research, so I became a part-time student in the Division of Sociomedical Sciences at the Joseph L. Mailman School of Public Health at Columbia University. The SMS team, a brilliant, caring group of professors who study history to guide and shape public policy, includes Amy Fairchild, PhD; David Rosner, PhD; Ronald Bayer, PhD; James Colgrove, PhD; Barron Lerner, MD, PhD; Nancy Stepan, PhD; and Sheila Rothman, PhD. They pushed me to delve deeper, broaden my outlook, and frame my questions within a historical context. (And Ron and David host great cocktail gatherings, and they kept inviting me during my leave of absence.)

You can't learn about childbirth from archives and doctors alone; you have to listen to the folks on the front line. There were probably at least 100 women who shared their birth stories with me. Pat Cody, a vivacious octogenerian, Leslie Myers (she taught me more about C-sections from her personal experiences than I ever learned from watching a few), Susan Helmrich, Wendy Kramer, Andrea Goldstein, Fran Howell, Priscilla Norton, Dorothy Tomashower, and the late Florence Wald were particularly helpful.

Al Filreis, Mingo Reynolds, and Jamie-lee Josselyn embraced me into the warm family of the Kelly Writer's House of the University of Pennsylvania and allowed me to give my first book reading from a draft of a chapter to the most welcoming audience. Nathan Kravis, MD, who runs the Richardson History of Psychiatry Research Seminar invited me to talk about psychogenic infertility in front of his distinguished members, who provided me with useful insights.

I have many friends who read chapters and offered much-needed encouraging words and friends who are probably sick of hearing about the book but were always willing to have cocktails and listen about this chapter or that research glitch—Janet Grillo, Karen Bank, Andrea Tone, Maria Romano, Sonia Best, Deirdre Depke, Sarah Key, Jennifer Stuart. Yes, Sarah Bank, the book is finally finished. Thanks to Jessica Baldwin, my British-based liaison, who helped me secure an article from a museum in Leeds. Sonya Lee, my children's wonderful piano teacher, was able over the phone to figure out which Mozart symphony accompanied the YouTube embryo clip. (I blasted the 55-second film and put my phone on speaker.) Mark Bloom gave me my first job in journalism and has been a mentor throughout my career, and although he believes that any story can be told in 800 words or less, I think he may like this rather lengthy one. Bob Barr, my editor when I worked in the London bureau of the Associated Press, probably triggered the first notions of this book when he encouraged me to investigate stories about the philosophy of medicine, not simply the breakthroughs. It was there I started to collect the anecdotes that eventually became part of this narrative. Nina Berman, a friend from journalism school and now a world-renowned photographer, spent hours taking my author photo, trying to make me look both scholarly and approachable. Alice and Tommy Tisch helped me to sail across the finish line.

And of course, this book would never have passed first base if it weren't for the Westside Little League Panthers. It was there, at the Riverside baseball fields, where I met my dear friends William Cohan and Deb Futter. Before we were fully introduced, Deb began to micromanage the process and insisted that I e-mail my proposal to Joy Harris, a literary agent, whom she called "a warm bath."

Joy Harris is my agent now and a good friend and best of all, my very own cheerleading squad. My editor, Jill Bialosky, at W. W. Norton read my book thoughtfully, carefully, and with a keen eye. She had wonderful suggestions and nixed all my title ideas (rightly so) and came up with *Get Me Out*. Her assistant, Adrienne Davich, responded with incredible calm to my onslaught of pestering e-mails. I owe a debt of gratitude to copy editor Janet Greenblatt for going over this manuscript with a fine-tooth comb; to Helen Yentus, for the jacket design; and to Francine Kass, the art director.

My mother, Ruth Hutter, listened to my paragraphs from her bed (after I woke her) and from her cell phone, even when she told me it was a bad time, and gave me honest feedback. My dad, Robert Hutter, was encouraging early on in the process, and although he suffers from Alzheimer's disease, I have to believe he is still more proud of me than anyone else that I've written a book. My sister, Edie, and my brother, Andrew, always seem to have more faith in my abilities than I sometimes have in myself.

My children, Jack, Martha, Joseph, and Eliza, have made every day more fun, even if they don't think my jokes are funny. Dexter, my gentle German Shepherd, sat loyally at my feet during the entire process. And to Stuart, who has heard every chapter at least seven times and learned to ignore my not-infrequent bouts of self-doubt: For nearly 30 years, he has made my life happier, tidier, and funnier (the kids laugh at his jokes). Who would have thought when we went to dinner the first night of freshman orientation that one day we would create our own history of childbirth.

Notes

Introduction

Chapter One

For a more in-depth account of childbirth attitudes in antiquity, see H. R. Lemay, *Women's Secrets: A Translation of Pseudo-Albertus Magnus' De Secretis Mulierum with Commentaries* (Albany: State University of New York Press, 1992).

3 **In 1591, Eufame Maclayne was burned:** S. Lurie, "Euphemia Maclean, Agnes Sampson and Pain Relief during Labor in 16th Century Edinburgh," *Anaesthesia* 59, no. 8 (2004): 834.

4 **Queen Victoria:** D. Caton, *What a Blessing She had Chloroform* (New Haven: Yale University Press, 1999), 54.

5 **Wert in 1522:** B. Hibbard, *The Obstetrician's Armamentarian* (San Anselmo, CA: Norman Publishing, 2000), 4.

6 **"groaning beer":** A. C. Banks, *Birth Chairs, Midwives, and Medicine* (Jackson: University Press of Mississippi, 1999), 50.

6 **gossips:** P. T. Ellison, *On Fertile Ground: A Natural History of Human Reproduction* (Cambridge: Harvard University Press, 2001), 55.

6 **advice about not having too much sex, but having some sex:** C. S. Rosenberg and Charles Rosenberg, "The Female Animal: Medical and Biological Views of Woman and Her Role in Nineteenth Century America," *The Journal of the American History* 60, no. 2 (1973): 333.

6 **"whores have so seldome children" . . . because "satiety gluts that womb":** I. Thomas, ed., *Culpeper's Book of Birth: A Seventeenth Century Guide to Having Lusty Children* (Devon, UK: Webb & Bower, 1985).

7 **"source of life":** L. Frieda, *Catherine de Medici Renaissance Queen of France* (New York: Fourth Estate, HarperCollins, 2003).

8 **"lay hold of the seed":** O. Temkin, trans., *Soranus' Gynecology* (Baltimore: Johns Hopkins University Press, 1956).

8 **normal uterus:** T. N. K. Raju, "Soranus of Ephesus: Who Was He and What Did He Do?" *Historical Review and Recent Advances in Neonatal and Perinatal Medicine*, Mead Johnson Nutritional Division (1980).

9 **sex during pregnancy, children "ill tempered, sickly, and short-lived":** J. H. M. Dye, *Painless Childbirth or Healthy Mothers & Healthy Children* (Silver Creek, NY: The Local Printing House, 1884).

12 **"and by that extent the pleasure may be mutually augmented":** E. Hobby, ed., *The Midwives Book or the Whole Art of Midwifery Discovered. Women Writers in English 1350–1850* (New York: Oxford University Press, 1999).

12 **large clitorises in women:** E. Hobby, ed., *The Midwives Book or the Whole Art of Midwifery Discovered. Women Writers in English 1350–1850* (New York: Oxford University Press, 1999), 40.

12 **"voluptuous itch":** I. Thomas, ed., *Culpeper's Book of Birth: A Seventeenth Century Guide to Having Lusty Children* (Devon, UK: Webb & Bower, 1985), 54.

12 **"loss of beauty"; "Man's Yard" and the advice about having odors waft through the body:** Francois Mauriceau, *The Diseases of Women with Child, and in the Child-Bed: As also the Best Means of Helping Them in Natural and Unnatural Labours, with Fit Remedies for the Several Indispositions of Newborn Babes, to Which is Prefix'd an Exact Description of the Parts of Generation in Women*, 7th ed., trans. H. T. Chamberlen (London: T. Cox and J. Clarke, 1763).

13 **wandering and suffocated womb:** L. Thompson, *The Wandering Womb: A Cultural History of Outrageous Beliefs about Women* (New York: Prometheus Books, 1999), 22– 37.

13 **"possess her roughly":** Helen R. Lemay, "Anthonius Guainerius and Medieval Gynecology," in *Women of the Medieval World*, ed. J. Kirshner and S. F. Wemple, 317–336 (New York: Basil Blackwell, 1985).

13 **"it often happens":** Helen R. Lemay, "Anthonius Guainerius and Medieval Gynecology," in *Women of the Medieval World*, ed. J. Kirshner and S. F. Wemple, 317–336 (New York: Basil Blackwell, 1985).

16 **"I'm talking about":** E. Shorter, *A History of Women's Bodies* (New York: Penguin Books, 1982), 35–36. The poem is Shorter's translation from Gustav Klein, ed., *Eucharius Rösslin's Rosengarten* (Munich, 1910), 8.

Chapter Two

18 **"To give you his character truly compleat":** B. Hibbard, *The Obstetrician's Armamentarian* (San Anselmo, CA: Norman Publishing, 2000).

18 **"when a man comes":** A. C. Banks, *Birth Chairs, Midwives, and Medicine* (Jackson: University Press of Mississippi, 1999), 25.

19 **in utero baptism:** K. Das, *Obstetric Forceps: Its History and Evolution* (1929; reprint, Leeds, UK: Medical Museum Publishing, 1993).

19 **Ayurvedic text:** K. Das, *Obstetric Forceps: Its History and Evolution* (1929; reprint, Leeds, UK: Medical Museum Publishing, 1993).

19 **"the woman is to be encouraged with hope by kind language":**

K. Das, *Obstetric Forceps: Its History and Evolution* (1929; reprint, Leeds, UK: Medical Museum Publishing, 1993).

19 **forceps definitions:** Michael J. O'Dowd and. Elliott E. Philipp, *The History of Obstetrics and Gynaecology* (New York: Parthenon Publishing Group, 1994); T. Parvin, "The Forceps," in *The Science and Art of Obstetrics* (Philadelphia: Lea and Brothers, 1895).

21 **"gaudy, frilly":** W. Radcliffe, *Milestones in Midwifery and The Secret Instrument (The Birth of the Midwifery Forceps)* (San Francisco: Norman Publishing, 1989); J. Mitford, *The American Way of Birth* (New York: Plume, Williams Abrams, 1993); B. Hibbard, *The Obstetrician's Armamentarian* (San Anselmo, CA: Norman Publishing, 2000).

22 **"grasping tightwads":** J. Mitford, *The American Way of Birth* (New York: Plume, Williams Abrams, 1993).

23 **William Smellie dressing up as a woman:** R. W. Wertz and Dorothy C. Wertz, *Lying-In: A History of Childbirth in America* (New Haven: Yale University Press, 1989), 41.

23 **the hoax, putting the instruments in a large box:** P. M. Dunn, "The Chamberlen Family (1560–1728) and Obstetric Forceps," *Archives of Disease Child Fetal Neonatal Edition* 81 (1999): F232.

24 **"I will not take apology":** Francois Mauriceau, *The Diseases of Women with Child, and in the Child-Bed: as also the Best Means of Helping Them in Natural and Unnatural Labours, with Fit Remedies for the Several Indispositions of Newborn Babes, to Which is Prefix'd an Exact Description of the Parts of Generation in Women*, 7th ed., trans. H. T. Chamberlen (London: T. Cox and J. Clarke, 1763), iii.

26 **"I am certain that no such thing as bringing a strange baby":** B. Hibbard, *The Obstetrician's Armamentarian* (San Anselmo, CA: Norman Publishing, 2000).

28 **the statue of Hugh:** The statue of Hugh is in the North Choir aisle, middle of the third bay, south side of Westminster Abbey. The detailed description, photo, and inscription was sent to me by the librarian of Westminster Abbey museum.

29 **"dangerous substitutes":** R. W. Wertz and Dorothy C. Wertz, *Lying-In: A History of Childbirth in America* (New Haven: Yale University Press, 1989), 90.

30 **poem about killing infants:** B. Hibbard, *The Obstetrician's Armamentarian* (San Anselmo, CA: Norman Publishing, 2000).

32 **forceps statistics :** R. Patel and D. Murphy, "Forceps Delivery in Modern Obstetrics Practice," *British Medical Journal* 328 (2004): 1302–1305.

33 **information about developing countries:** P. E. Bailey, "The Disappearing Art of Instrumental Delivery: Time to Reverse the Trend," *International Journal of Gynecology and Obstetrics* 91 (2005): 89–96; personal interview with P. Bailey.

34 **"Though their assumptions":** C. M. Scholten, *Childbearing in American Society: 1650–1850* (New York: New York University Press, 1985), 39.

Chapter Three

36 **no longer "fit to be around other people":** C. M. Scholten, *Childbearing in American Society: 1650–1850* (New York: New York University Press, 1985).

37 **"My son, I must confess":** J. M. Sims, *The Story of My Life* (New York: D. Appleton and Co., 1898; reprint, Whitefish, MT: Kessinger's Publishing Rare Reprints, 2007), 116.

37 **kicked in:** C. Rosenberg, "Belief and Ritual in Antebellum Medical Therapeutics," in *Major Problems in the History of American Medicine and Public Health*, ed. J. H. Warner and J. A. Tighe (New York: Houghton Mifflin, 2001) 110.

38 **"messy and unscientific":** S. Harris, *Woman's Surgeon: The Life Story of J. Marion Sims* (New York: The MacMillan Co., 1950), 28.

38 **a British blacksmith and grocer:** I. Loudon, *Death in Childbirth: An International Study of Maternal Care and Maternal Mortality 1800–1950* (Oxford: Clarendon Press, 1992), 174–175.

39 **"very clannish":** J. M. Sims, *The Story of My Life* (New York: D. Appleton and Co., 1898; reprint, Whitefish, MT: Kessinger's Publishing Rare Reprints, 2007), 207.

40 **"I am relieved":** J. M. Sims, *The Story of My Life* (New York: D. Appleton and Co., 1898; reprint, Whitefish, MT: Kessinger's Publishing Rare Reprints, 2007), 233.

40 **"greatest discoveries of the day":** J. M. Sims, *The Story of My Life* (New York: D. Appleton and Co., 1898; reprint, Whitefish, MT: Kessinger's Publishing Rare Reprints, 2007), 235.

42 **"as no man had ever seen before":** J. M. Sims, *The Story of My Life* (New York: D. Appleton and Co., 1898; reprint, Whitefish, MT: Kessinger's Publishing Rare Reprints, 2007), 235.

43 **"unfitted me for the almost":** J. M. Sims, *The Story of My Life* (New York: D. Appleton and Co., 1898; reprint, Whitefish, MT: Kessinger's Publishing Rare Reprints, 2007), 235.

46 **"was a rapture in his work":** S. Harris, *Woman's Surgeon: The Life Story of J. Marion Sims* (New York: The MacMillan Co., 1950), xx.

46 **"Among the greater luminaries":** S. Harris, *Woman's Surgeon: The Life Story of J. Marion Sims* (New York: The MacMillan Co., 1950), 25.

46 **"How blessed and sweet":** J. H. Rice, Jr., eulogy, *James Marion Sims Memorial* (Columbia, SC: South Carolina Medical Association, 1929), 26.

47 **"Sims was a man of noble character":** J. P. Marr, *James Marion Sims: The Founder of the Woman's Hospital in the State of New York* (New York: The Woman's Hospital, 1949), last page.

48 **"Hideous as the accounts":** C. M. de Costa, "James Marion Sims: Some Speculations and a New Position," *Medical Journal of Australia* 178 (2003): 663.

48 **"dedicated and conscientious":** L. L. Wall, "The Medical Ethics of Dr. J. Marion Sims: A Fresh Look at the Historical Record," *Journal of Medical Ethics* 32, no. 6 (2006): 350.

48 **"ethical balance sheet":** H. Washington, *Medical Apartheid: The Dark History of Medical Experimentation on Black Americans from Colonial Times to the Present* (New York: Doubleday, 2006).

48 **"indomitable courage":** John H. Warner and Janet A. Tighe, eds., *Major Problems in the History of American Medicine and Public Health*, chaps. 4–6 (Boston: Houghton Mifflin, 2001), 123.

Chapter Four

For a thorough analysis of the history of childbed fever and maternal mortality, see I. Loudon, *Death in Childbirth: An International Study of Maternal Care and Maternal Mortality 1800–1950* (Oxford: Clarendon Press, 1992).

52 **Lydia Pinkham:** "What Lydia E. Pinkham's Vegetable Compound Is Doing for Women," advertisement in the *Boston Daily Globe*, 1 April 1900, 9.

54 **"like a never-silent church bell":** F. C. Irving, *Safe Deliverance* (Boston: Houghton Mifflin, 1942).

54 **constipation and wafts of cold air:** J. Gelis, *History of Childbirth: Fertility, Pregnancy, and Birth in Early Modern Europe* (Boston: Northeastern University Press, 1991).

57   **Dr. McLean: Van Hoosen's memoir, where she describes him as charismatic and handsome:** B. V. Hoosen, *Petticoat Surgeon* (Chicago: People's Book Club, 1947).

58   **excitable or external causes:** M. Worboys, *Spreading Germs: Disease Theories and Medical Practice in Britain 1865–1900* (New York: Cambridge University Press, 2000).

60   **maternal mortality, Baker's statement:** Shockingly so, in 2008, the Centers for Disease Control and Prevention's National Center for Health Statistics ranked the United States 29th globally in infant mortality. This, of course, is different from issues of maternal mortality, but does point to some concerns about childbirth and the care of newborns. The findings were reported in the February 2009 issue of *The Nation's Health*, the newspaper of the American Public Health Association. The investigators say that one of the factors influencing our abysmal ranking may be the incidence of premature births—in 2005, 68 percent of the deaths occurred among premature infants. The researchers wonder whether health disparities, where some mothers are not receiving proper care or nutrition during pregnancy, is affecting the childbirth outcomes. They did not mention the use of the new reproductive technologies, which increase the number of multiples and, in turn, increase the risk of prematurity.

60   **"infecting or by the spread":** H. Vineberg, "The Treatment of Puerperal Sepsis," *American Journal of Obstetrics* 48 (September 1903).

61   **"from the accoucheur":** S. Marx, "The Bacteriology of the Puerperal Uterus: Its Relation to the Treatment of the Parturient State," *American Journal of Obstetrics* 48, no. 3 (1903): 301–323.

Chapter Five

Much of the information and newsclippings for this chapter was gleaned from the Scrapbooks of the Lying-In Hospital: 1890–1937 (Medical Center Archives of New York-Presbyterian/Weill Cornell, New York, NY).

64   **New Jersey letter:** Scrapbooks of the Lying-In Hospital: 1890–1937 (Medical Center Archives of New York-Presbyterian/Weill Cornell, New York, NY).

64   **If a woman died:** W. Radcliffe, *Milestones in Midwifery and The Secret Instrument (The Birth of the Midwifery Forceps)* (San Francisco: Norman Publishing, 1989).

64   **"You've nothing on me":** F. C. Irving, *Safe Deliverance* (Boston: Houghton Mifflin, 1942).

65 **In 1990, 5 percent of women . . . 1930s, 1960s:** R. W. Wertz and Dorothy C. Wertz, "Birth in the Hospital," in *Lying-In: A History of Childbirth in America*, chap. 5 (New Haven: Yale University Press, 1989).

66 **Hosack history:** J. A. Harrar, *The Story of the Lying-In Hospital of the City of New York* (New York: The Society of the Lying-In, 1938).

67 **the term *hospitalism*:** C. E. Rosenberg, *The Care of Strangers: The Rise of America's Hospital System* (New York: Basic Books, 1987).

68 **3-hour ride from home to hospital:** D. Rosner, *A Once Charitable Enterprise: Hospitals and Health Care in Brooklyn and New York 1885–1915* (New York: Cambridge University Press, 1982), 14.

69 **Overall, some 591 women:** S. Lambert and H. Painter, "Fever in the Puerperal Woman," *Dispensary Report*, 1893. Scrapbooks of the Lying-In Hospital: 1890–1937 (Medical Center Archives of New York-Presbyterian/Weill Cornell, New York, NY).

69 **"squalid, filthy, and wretched":** Scrapbooks of the Lying-In Hospital 1890–1937 (Medical Center Archives of New York-Presbyterian/Weill Cornell, New York, NY).

71 **"sort of woman":** Scrapbooks of the Lying-In Hospital 1890–1937 (Medical Center Archives of New York-Presbyterian/Weill Cornell, New York, NY).

71 **"if physicians insisted on procedures":** J. W. Leavitt, "Only a Woman Can Know: The Role of Gender in the Birthing Room," chap. 4 in *Brought to Bed: Childbearing in America, 1750–1950* (New York: Oxford University Press, 1986).

72 **"the destitute":** B. V. Hoosen, *Petticoat Surgeon* (Chicago: People's Book Club, 1947), 72.

72 **bake sale:** D. Rosner, *A Once Charitable Enterprise: Hospitals and Health Care in Brooklyn and New York 1885–1915* (New York: Cambridge University Press, 1982).

76 **"distinction between hospital and home":** J. B. DeLee, *The Principles and Practice of Obstetrics* (Philadelphia: W. B. Saunders, 1920), 280.

Chapter Six

The bulk of this chapter was gleaned from newspaper articles in the Eliza Ransom Papers, microfilm M-61, 12 Volume 2, Scrapbook on Twilight Sleep 1914–1915, at the Schlesinger Library, Radcliffe Institute for Advanced Studies, Harvard University.

79 **"night of my confinement":** M. Tracy and M. Boyd, *Painless Childbirth: A General Survey of All the Painless Methods, with Special Stress on "Twilight Sleep" and Its Extension to America* (New York: Frederick A. Stokes Company, 1915).

79 **bobbing hair:** S. Coontz, *Marriage, A History* (New York: Viking, 2005), 197.

80 **Bryant's clothes:** M. Feldberg, ed., *Blessings of Freedom: Chapters in American Jewish History* (Hoboken, NJ: KTAV, 2002); also available at www.ajhs.org.

81 **freshman 15:** J. Zeitz, *Flapper: A Madcap Story of Sex, Style, Celebrity and the Women Who Made America Modern* (New York: Three Rivers Press, 2006).

81 **"What more charming sight":** A. B. Stockman, *Tokology* (New York: R. F. Fenno & Company, 1911).

81 **More than a half million:** G. Collins, *America's Women: Four Hundred Years of Dolls, Drudges, Helpmates, and Heroines* (New York: William Morrow, 2003), 311.

81 **Sex Side of Life:** "Mrs. Dennett Guilty in Sex Booklet Case," *New York Times*, 29 April 1929, 31.

82 **"The women of America are demanding":** M. Tracy and M. Boyd, *Painless Childbirth: A General Survey of All the Painless Methods, with Special Stress on "Twilight Sleep" and Its Extension to America* (New York: Frederick A. Stokes Company, 1915).

82 **membership in the Twilight Sleep Association:** J. Kaplan, *When the Astors Owned New York* (New York: Plume, 2006), 3.

83 **Morton, dentist:** D. Caton, *What a Blessing She Had Chloroform* (New Haven: Yale University Press, 1999), 4.

84 **Longfellow's wife:** D. Sanghavi, "The Mother Lode of Pain; Some Women Insist on a Drug-Free Childbirth—Even Though It Might Be Agonizing—While Others Opt for a Numbing Epidural. Is This All Part of the Simmering Debate Between Natural and Modern Medicine, or Are Some Women Embracing Labor Pain for a More Heroic Cause?" *Boston Globe Magazine*, 23 July 2006, 18–21, 28–29.

84 **drugs used to speed labor:** J. W. Leavitt, "Birthing and Anesthesia: The Debate Over Twilight Sleep," *Signs: Journal of Women in Culture and Society* 6, no.1 (1980): 147–164.

85 **"modern woman . . . responds":** M. Tracy and C. Leupp, "Painless Childbirth," *McClure's Magazine*, vol. 43, 1914, 22.

85 **Better Babies Movement:** J. W. Leavitt, "Birthing and Anesthesia: The Debate Over Twilight Sleep," *Signs: Journal of Women in Culture and Society* 6,

no. 1 (1980): 147–164; and A. S. Richardson, "What Every Mother Wants to Know about Her Baby," *Atlantic Constitution*, 4 August 1914, 12.

86 **"the propagation of life":** W. Hutchinson, "The Kind of Woman Who Ought to Have Babies." *Washington Post*, 16 July 1916, MT8.

86 **"Painless Childbirth":** M. Tracy and C. Leupp, "Painless Childbirth," *McClure's Magazine*, vol. 43, 1914.

86 **"Lifting the Curse of Eve":** D. D. Bromley, "Lifting the Curse of Eve," *Woman's Journal* 1 (October 1927): 8–10, 36–37.

86 **"Twilight Sleep Is Necessity":** "Twilight Sleep Is Necessity, Not Luxury," *International News Service*, 31 May 1916.

86 **"Drug Boon to Women":** "Drug Boon to Women: New Treatment for Childbirth Called Medical Mercy," *Washington Post*, 27 August 1914, 4.

87 **oligopnea:** J. O. Polak, "A Study of Twilight Sleep with a Critical Analysis of the Cases at the Long Island College Hospital." *New York Medical Journal* (February 13, 1915).

87 **Hazel Harris:** Hazel Harris, *New York Times*, 28 May 1914.

88 **ode to twilight sleep:** E. Wolff, "The Bridge of Dreams," *New York Times*, 28 May 1914, 12.

90 **"pressure from the women themselves":** C. Leupp and B. J. Henrick, "Twilight Sleep in America," *McClure's Magazine*, April 1915, 25–36.

91 **"Although she is dead":** "Drug Boon to Women: New Treatment for Childbirth Called Medical Mercy," *Washington Post*, 27 August 1914, 4.

91 **Anti–Twilight Sleep Association:** "Anti–Twilight Sleep Association to Fight Twilight Sleep: A Brooklyn Woman to Start Association to Oppose the Treatment," *New York Times*, 13 August 1915, 5.

91 **"enthusiastic hysteria":** J. Hamilton, "The Twilight Sleep: Why Physicians Are Conservative and Why Women May Not Expect Too Much from It," *The Evening Sun*, 23 April 1914.

Chapter Seven

95 **the information about Sylvia:** E. Jacobson, "A Case of Sterility," *Psychoanalytic Quarterly* 15, no. 3 (1946): 330–350. Reprinted with permission from the Archives and Special Collections, A. C. Long Health Sciences Library, Columbia University. The articles and correspondence are part of the extensive collection of Dr. Viola Bernard, subsequently referred to as VWB. The life and work of Dr. Viola Bernard (1907–1998) serve as

the bridge to join these two seemingly disparate fields of gynecology and psychiatry. Dr. Bernard's career, her obsession, was exploring fertility and the family—indeed, what constitutes a properly functioning family. Born to a wealthy New York City family, Dr. Bernard graduated from New York University in 1933, Cornell Medical School in 1936, and was one of three women trained in both neurology and psychiatry. She founded Columbia University's Division of Community and Social Psychiatry, which existed from 1956 to 1969. She was a member of the New York Psychoanalytic Institute and fellow of the American Psychoanalytic Association. Of her many studies and collaborations, she interviewed and analyzed couples pre- and post-conception and pre- and post-adoption. She served as the psychiatric consultant to the Lois Wyse Services, an adoption agency.

96 **your thoughts make you sterile:** D. W. Orr, "Pregnancy Following Decision to Adopt," *Psychosomatic Medicine* 3, no. 4 (1941): 441–446, in Viola W. Bernard Archives; E. Jacobson, "A Case of Sterility," *Psychoanalytic Quarterly* 15, no. 3 (1946): 330–350, in Viola W. Bernard Archives. The Jacobson article is one of many during that time that included case analyses of infertile women. A thorough collection of studies by other psychoanalysts is included in the Viola W. Bernard Archives, several of which will be cited below. For instance, Dr. Bernard has patient records and drafts of articles from her colleague, Dr. Kenneth Kelley, who also practiced at the Departments of Psychiatry and Medicine, Columbia University. Dr. Bernard also has case studies from Dr. Ruth Moulton, who studied dreams of infertile women.

97 **Psychogenic infertility . . . a theory without any proof whatsoever:** The Viola W. Bernard Archives also contain a wealth of correspondence demonstrating the growing popularity of the notion of psychosomatic infertility, with such leaders as Dr. John Rock, who later became a pioneering investigator in the oral contraceptive; Dr. Isador Clinton Rubin, who devised a routine diagnostic test for infertility; and Dr. Sophia Kleegman, one of the first female infertility specialists. Roberto A. Volta, writing from Buenos Aires, Argentina, September 28, 1950: Dr. Volta writes to Viola Bernard that he spoke about her work on psychogenic infertility at a meeting in Buenos Aires and has been inundated with requests for her papers. Most of the scientific articles are in psychiatric journals and journals for psychoanalysts, but several articles were published in journals for a nonpsychiatric audience, such as obstetric journals and one citing in *Lancet:* Therese Benedek,

George Ham, et al., "Some Emotional Factors in Infertility," *Psychosomatic Medicine* 15, no. 5 (1953): 485–498, from the Viola W. Bernard Archives; I. C. Rubin, "Diagnostic Procedures in the Investigation of Sterility in the Female: Evalution of Their Clinical Importance," in *Collected Papers of Dr. I. C. Rubin 1910–1954*, from the Viola W. Bernard Archives; Douglass W. Orr, "Pregnancy Following the Decision to Adopt," *Psychosomatic Medicine* 3, no. 4 (October 1941): 441–446, from the Viola W. Bernard Archives; Frederick W. Dershimer, "Influence of Mental Attitudes in Childbearing," *American Journal of Obstetrics and Gynecology* 31, no. 3 (March 1936): 444; Ernst Jones, "Psychology and Childbirth," *Lancet* 242 (June 6, 1942): 695–696. (This brief overview describes how neurotic attitudes, beginning in toddlerhood, can impact fertility, labor, and childbirth.); Kenneth Kelley, "Sterility in the Female," *Psychosomatic Medicine* 4, no. 2 (1942).

97   **"psychic menorrhagia," organotherapy:** J. M. Miller, "Psychogenic Menorrhagia," *Medical Journal and Record* (July 15, 1931):84–86.

99   *pituitary* **is Latin for "slime":** R. Porter, *The Greatest Benefit to Mankind* (New York: W. W. Norton, 1997), 565.

100   **keynote address:** James E. King, "Presidential Address: American Association of Obstetricians, Gynecologists, and Abdominal Surgeons," *American Journal of Obstetrics and Gynecology* 39, no. 2 (February 1939): 180–181.

101   **Planned Parenthood meeting:** V. Bernard, "Abstract of Talk at the National Conference of Social Work for Planned Parenthood on Psychiatric Aspects of Infertility in Women," 15 April 1947, from Box 58, the Archives and Special Collections, A. C. Long Health Sciences Library, Columbia University.

103   **Consumer magazines:** Advertisements, many by Parke, Davis, and Co., were published in consumer publications. One, for instance, called "Just a Bundle of Nerves," directs women who are irritable and exhausted to consult their physician for drug therapy (*Hygiea*, April 1940). Among other articles: L. Pruette, "Critical Days for Mother," *Parents* 35, 15 May 1940; L. M. Miller, "Changing Life Sensibly: New Hormone Treatments for Menopause," *Independent Woman*, 18 September 1939, 297, also in *Reader's Digest* vol. 35, October 1939, 101–103; M. C. Miller, "Facts about Menopause," *Hygiea* (August 1940): 692–694; J. H. Kenyon, "Tired Mother," *Good Housekeeping* vol. 110, January 1940, 685–688. Smith-Rosenberg has referred to the New Woman, who was seeking a career yet also pushed to rear emotionally balanced children.

104 **"re-regulating of neuro":** Viola W. Bernard, Draft of a Talk, dated 1942 in the Viola W. Bernard Archives. The files contain many drafts of planned talks, some of which may not have been presented due to postponements of meetings, but the papers, particularly the notes she scribbles to herself, reveal a lot about medical acceptance of psychogenic infertility, as she has several remarks about making a convincing argument and deleting anything that may seem too outlandish for a nonpsychiatric audience.

104 **NIH study:** L. McLaughlin, *The Pill, John Rock, and the Church* (Boston: Little, Brown and Company, 1982), 49.

104 **For the most part:** I. C. Rubin, *Collected Papers of I. C. Rubin 1910–1954* (bound reprints of papers, Columbia Health Sciences Library). "Dr. Rubin (1883–1958) is considered by some the father of modern scientific investigation of infertility": Michael O'Dowd and Elliott E. Philipp, *The History of Obstetrics and Gynaecology* (New York: Parthenon Publishing Group, 1994). Dr. Rubin graduated from the College of Physicians and Surgeons and continued his studies in gynecologic pathology in Austria and Germany. Besides his Rubin test for fallopian tube patency, he also investigated carcinoma of the cervix, trying to investigate minimal disease. By 1954, he had written more than 130 scientific papers, primarily dealing with infertility. Dr. Rubin was internationally recognized, receiving many honorary degrees, including one from the Royal College of Obstetricians and Gynecologists. He was also president of the American Gynecological Society in 1955. His nod of approval to the psychiatrists and psychoanalysts studying infertility was immensely helpful. His papers that refer to psychiatric factors include: "Diagnostic Procedures in the Investigation of Sterility in the Female: Evaluation of Their Clinical Importance" (presented at the afternoon lunch series, New York Academy of Medicine, 7 February 1947), in I. C. Rubin, *Collected Papers*. He cites similar views in a lecture to the 29th annual meeting of the Lake Keuka Medical and Surgical Association, Keuka, New York, 12–13 July 1928). Among some of the papers and talks with references to psychiatric influences are: "Ovarian Hypofunction, Habitually Delayed and Scanty Menstruation in Relation to Sterility and Lowered Fertility" (presented at the 54th annual meeting of the American gynecological society, Old Point Comfort, VA, 20–22 May 1929); and "Thirty Years of Progress in Treating Infertility," *Fertility and Sterility* 1 (September 1950): 5, 389–406. This article is a great source of historical information about fertility treatments.

104 **Articles that sounded like:** K. A. Meninger, "Somatic Correlations with Unconscious Repudiation of Fertility in Women," *Journal of Nervous and Mental Diseases* 89 (April 1939): 514–527, also appearing in *Bulletin of the Meninger Clinic* 3 (July 1939): 106– 121; Therese Benedek and Boris B. Rubenstein, "The Correlations Between Ovarian Activity and Psychodynamic Processes: II. The Menstrual Phase," *Psychosomatic Medicine* 1, no. 4 (October 1939): 461–485.

104 **"As her attitudes towards herself changed, her pelvic physiology":** D. W. Orr, "Pregnancy Following Decision to Adopt," *Psychosomatic Medicine* 3, no. 4 (1941): 441–446.

104 **25 of 29 infertile patients with mental illness:** Kenneth Kelley, "Sterility in the Female," *Psychosomatic Medicine* 4, no. 2 (1942); from the Viola W. Bernard Archives.

105 **"hollow triumph":** T. Benedek, "Infertility as a Psychosomatic Defense," *Fertility and Sterility* 3, no. 6 (1952): 527–537.

106 **Domar:** Alice Domar, *Conquering Infertility: Dr. Alice Domar's Mind/Body Guide to Enhancing Fertility and Coping with Infertility* (New York: Penguin, 2004); personal interviews.

107 **Berga:** S. Berga, "Psychiatry and Reproductive Medicine," in *Kaplan & Sadock's Comprehensive Textbook of Psychiatry*, 7th ed., vol. 2, ed. Benjamin Sadock and Virginia Sadock, 1935–1952 (Philadelphia: Lippincott Williams & Wilkins, 2000); S. Berga, "Recovery of Ovarian Activity in Women with Functional Hypothalamic Anmenorrhea Who Were Treated with Cognitive Behavior Therapy," *Fertility and Sterility* 80, no. 4 (2003): 976–980; personal interviews.

108 **Rosner:** Personal interview.

108 **Bing:** Personal interview.

Chapter Eight

109 **Bing:** Several personal interviews in the fall and spring of 2006–2007.

111 **"Happy childbirth":** "Baby Born and No Pain at All: Author of Childbirth without Fear Dr. Grantly Read Visiting America," *The Herald* (Bridgeport), 19 January 1947; from the Herbert Thoms Collection, Historical Library, Cushing/Whitney Medical Library, Yale University.

111 **"as if it were another limb":** F. Kartchner, "A Study of the Emotional Reactions during Labor," *American Journal of Obstetrics and Gynecology* 60, no. 1 (1950): 19–29.

111 **hypnosis:** William Kroger, *Childbirth with Hypnosis* (New York: Doubleday, 1961).

112 **"A tense woman is closing the door against her baby":** G. Dick-Read, *The Natural Childbirth Primer* (New York: Harper & Brothers, 1955).

113 **the Soviet-Russian and Communistic Chinese:** Personal correspondence; from the Herbert Thoms Collection, Historical Library, Cushing/Whitney Medical Library, Yale University.

115 **internal memos of Maternity Center Association:** From the Herbert Thoms Collection, Historical Library, Cushing/Whitney Medical Library, Yale University.

116 **Christian organizations pushed natural childbirth:** R. W. Wertz and Dorothy C. Wertz, *Lying-In: A History of Childbirth in America* (New Haven: Yale University Press, 1989), 157.

116 **"Woman is made primarily in order that children might be born into this world . . . yet" and the following speech:** A talk by Dr. Grantly Dick-Read to the New York Academy of Medicine, 17 January 1947; from the Herbert Thoms Collection, Historical Library, Cushing/Whitney Medical Library, Yale University.

118 **"selfish, introverted woman":** C. Kemp, "British Doctor Claims Suffering Results from Physical Derangement or Misunderstanding on Part of the Mother," *Courier Journal* (Louisville, KY), February 23, 1947; from the Herbert Thoms Collection, Historical Library, Cushing/Whitney Medical Library, Yale University.

119 **Yale Dames:** This information is gleaned from talking to a group of women who were part of the Yale Dames or friends with them, including interviews with Dr. Morris Wessel, a Yale University pediatrician who was friends with the "dames"; Dr. Wessel's wife, Irmgard Wessel, who also remembers them well; Priscilla Norton; and the late Florence Wald, who had been dean of the school of nursing at Yale and founded the American hospice movement.

121 **Goodrich study:** F. Goodrich, Jr., and Herbert Thoms, "A Clinical Study of Natural Childbirth: A Preliminary Report from a Teaching Ward Service," *American Journal of Obstetrics and Gynecology* 56, no. 5 (1948): 875–883; Frederick W. Goodrich, Jr., "The Theory and Practice of Natural Childbirth," *Yale Journal of Biology and Medicine* 25, no. 6 (June 1953): 529–534. In the latter article, Goodrich expresses his attitudes and those of many at Yale when he says that "natural childbirth is basically a system of intellectual, emotional, and physical preparation for childbirth to the end that

mothers may enjoy a healthier and happier pregnancy and delivery. . . . I believe the day is past when we could expect blind obedience from our patients." He says that this does not mean minimal drugs or a certain amount; it means that the patient is educated, that "the choice of how much anesthesia to be given is thus up to the patient." Goodrich says that "natural childbirth is a philosophy of *total* [his italics] preparation in which the mother is intellectually and emotionally prepared as well. . . . As a consequence . . . more mothers need less help from drugs and anesthetics than has heretofore been the case."

122 **psychiatric dangers lurked:** L. Freedman and V. M. Ferguson, "The Question of 'Painless Childbirth' in Primitive Cultures," *The American Journal of OrthoPsychiatry* 20, no. 2 (1950).

122 **Other hospitals followed suit:** E. B. Jackson, "Mothers and Babies Together," *Parents Magazine* vol. 22, 1947, 18–19, 146–149; E. B. Jackson, "Rooming In Gives Baby a Good Start," *The Child* 12 (1948): 162–165; E. B. Jackson, "The Rooming-In Plan for Mothers and Infants," *American Journal of Obstetrics and Gynecology* 65 (1948): 707–711.

123 **The press was all over this story:** G. Schultz, "Cruelty on the Maternity Wards," *Ladies Home Journal*, 1958, 44–45, 152, 155; Malvina Lindsay, "Modern Trek to Nature," *Washington Post*, 27 January 1949, 10; several newspaper clippings from the Herbert Thoms Collection, Historical Library, Cushing/Whitney Medical Library, Yale University.

124 **Maternal mortality dropped 70 percent:** "CDC on Infant and Maternal Mortality in the United States: 1900–99," *Population and Development Review* 25, no. 4 (1999): 821–826.

124 **Bernice Levine:** Personal interview.

126 **Studdiford, "Tale of Two Cities":** William Studdiford, "Childbirth Difference Is Tale of Two Cities," *West Hartford News* (CT), 19 January 1950; from the Herbert Thoms Collection, Historical Library, Cushing/Whitney Medical Library, Yale University.

127 **Dorothy Thomashower:** Personal interview.

Chapter Nine

The information in this chapter is based on several interviews with DES activists, including Fran Howel, desaction@comcast.net; Pat Cody, with several e-mail interviews and a copy of her detailed self-published manuscript about the history of DES; and Andrea Goldstein, who is a DES mother and has kept her own personal archives of DES articles.

130 **"You can't imagine what it's like":** D. Bryant, "The DES Odyssey of Pat Cody," *San Francisco Sunday Examiner and Chronicle*, 18 March 1979, 17–22.

131 **Pitkin:** Personal interview.

131 **doctors noticed newborns with traces of lead:** D. Caton, *What a Blessing She Had Chloroform: The Medical and Social Response to the Pain of Childbearing from 1800 to Present* (New Haven, CT: Yale University Press, 1999), 77.

133 **"least undesirable day":** B. Asbell, *The Pill: A Biography of the Drug That Changed the World* (New York: Random House, 1995), 18.

133 **concoction of hormones that juggled our hormones in a slightly different way:** B. Asbell, *The Pill: A Biography of the Drug That Changed the World* (New York: Random House, 1995), 171; L. McLaughlin, *The Pill, John Rock, and the Church* (Boston: Little, Brown and Company, 1982), 139.

133 **Thalidomide is a safe drug for pregnant rats:** T. D. Stephens and Rock Brynner, *Dark Remedy: The Impact of Thalidomide and Its Revival as a Vital Medicine* (New York: Perseus Publishing, 2001).

133 **restoring the body to its original balance:** M. Gladwell, "John Rock's Error," *New Yorker*, 13 March 2000, 52–63.

134 **Sir Charles Dodds trimmed two of estrogen's:** E. Shorr, Frank Robinson, and George N. Papanicolaou, "A Clinical Study of the Synthetic Estrogen Stilbestrol," *Journal of the American Medical Association* 113, no. 26 (1939): 2312–2318.

135 **better understanding of the proper functioning of such drugs:** P. N. Leech, "Preliminary Report of the Council: Stilbestrol," *Journal of the American Medical Association* 113, no. 26 (1939): 2312.

135 **"The possibility of carcinoma":** P. Cody, *The Silent Epidemic* (Berkeley, California, 2006), 185.

136 **off-label uses:** R. Apfel and S. M. Fisher, *To Do No Harm* (New Haven: Yale University Press, 1984); R. J. Apfel, "Psychoanalysis and Infertility," *International Journal of Psychoanalysis* 83 (2002): 85–104.

136 **Many women were getting a lifetime's worth of estrogen:** C. L. Orenberg, *DES: The Complete Story* (New York: St. Martin's Press, 1981).

136 **"roughly the equivalent of 500,000 of the current low-dose birth control pills":** P. Cody, *The Silent Epidemic* (Berkeley, California, 2006), 11.

136 **correspondence, Smith:** George and Olive Watkins Smith Papers [H MS c68] at the Center for the History of Medicine, Countway Library of Medicine, 1950, 1951.

140 **Australian doctor:** T. Lovegrove, letter to Olive Smith, from Wonghan Hills, West Australia, 25 November 1965.

142 **"Scientific progress is being made in discovering ways":** M. Fishbein, "Too Many Miscarriages," *Woman's Home Companion*, vol. 71, 1944, 20, 37.

143 **"but I'd like to feel that I'm doing everything":** C. Friess, "What We Know about Miscarriage," *Woman's Home Companion*, vol. 82, 1955, 4–6.

143 **Between 1966 and 1969, a rare form of vaginal cancer called adenocarcinoma:** A. L. Herbst, Howard Ulfelder, and David Poskanzer, "Adenocarcinoma of the Vagina: Association of Maternal Stilbestrol Therapy with Tumor Appearance in Young Women," *New England Journal of Medicine* 284, no.16 (1971): 878–881; personal interviews.

145 **Dr. Peter Greenwald:** E-mail correspondence.

147 **On April 23, 1971, the news of DES dangers hit all leading newspapers:** "FDA Warning on Synthetic Estrogen" (*Los Angeles Times*), "Girls Cancer Laid to Mother's Drug" (*Washington Post*), "Rare Cancer Type Linked to A Drug (*New York Times*).

147 **Susan Helmrich:** Personal interviews.

149 **In the early 1980s, a few lawyers fought for legal changes:** P. Cody, *The Silent Epidemic* (Berkeley, California, 2006), 96.

150 **Anna Quindlen:** Anna Quindlen, *New York Times*, 9 May 1993, section 4, 15; also appears in P. Cody, *The Silent Epidemic* (Berkeley, California, 2006), 116.

Chapter Ten

153 **Haseltine:** Personal interview.

153 **90 percent of obstetricians were men:** E. Frank, John Rock, and Danielle Sara, "Characteristics of Female Obstetricians-Gynecologists in the United States," *Obstetrics and Gynecology* 94, no. 5 (1999): 659–665.

154 **one in three babies were born surgically:** F. Menacker, "Trends in Cesarean Rates for First Births and Repeat Cesarean Rates for Low-Risk Women: United States, 1990–2003," *National Vital Studies Reports* 54, no. 4 (2005): 1–12; available at www.cdc.gov/nchs.

155 **early history of Caesar and caesarean births:** Michael J. O'Dowd and

Elliott E. Philipp, *The History of Obstetrics and Gynaecology* (New York: Parthenon Publishing Group, 1994), 157–163.

157 **priests reference and "her breath leaves no traces on a mirror":** E. Shorter, *A History of Women's Bodies* (New York: Penguin Books, 1982), 160.

157 **Nufer:** Michael J. O'Dowd and Elliott E. Philipp, *The History of Obstetrics and Gynaecology* (New York: Parthenon Publishing Group, 1994), 161.

158 **Long Island C-section:** H. Speert, *Obstetrics and Gynecology in America* (Baltimore: Waverly Press, 1980), 150.

158 **Turkish woman:** H. Churchill, *Caesarean Birth: Experience, Practice and History* (Cheshire, England: Books for Midwives Press, an imprint of Elsevier Limited, 1997), 30.

159 **"a large dose of laudanum":** H. Speert, *Obstetrics and Gynecology in America* (Baltimore: Waverly Press, 1980), 152.

159 **Max Sanger:** R. W. Wertz and Dorothy C. Wertz, *Lying-In: A History of Childbirth in America* (New Haven: Yale University Press, 1989), 139.

160 **Mrs. Payne, an African-American Virginian … C-section in front of several doctors:** T. L. Savitt, *Medicine and Slavery: The Diseases and Health Care of Blacks in Antebellum Virginia* (Chicago: University of Illinois Press, 1978), 118.

160 **"The Effect of Overcivilization on Maternity":** F. S. Newell, "The Effect of Overcivilization on Maternity," *American Journal of the Medical Sciences* 136 (1908): 533–541; J. W. Leavitt, *Brought to Bed: Childbearing in America, 1750–1950* (New York: Oxford University Press, 1986), 68.

161 **"The women of the large urban centers…at great risk to his own practice":** R. S. Hooker, *Maternal Mortality in New York City: A Study of All Puerperal Deaths 1930–1932* (New York: The Commonwealth Fund, 1933), 113.

161 **1933 maternal mortality study:** R. S. Hooker, *Maternal Mortality in New York City: A Study of All Puerperal Deaths 1930–1932* (New York: The Commonwealth Fund, 1933), 138.

162 **Leslie:** Personal interview.

162 **"Once a Caesarean always a Caesarean":** R. W. Wertz and Dorothy C. Wertz, *Lying-In: A History of Childbirth in America* (New Haven: Yale University Press, 1989), 139, 263.

163 **33,000-woman C-section study:** Mark B. Landon, John C. Hauth, Kenneth J. Leveno, Catherine Y. Spong, et al., "Maternal and Perinatal Out-

comes Associated with a Trial of Labor after Prior Cesarean Delivery," *The New England Journal of Medicine* 351, no. 25 (2004): 2587; Michael F. Greene, "Vaginal Birth after Cesarean Revisited," *New England Journal of Medicine* 351, no. 25 (2004): 2647.

165 **"We took this tool, electronic fetal monitoring":** Personal interview with Naftolin.

165 **A large meta-analysis of nine randomized controlled trials, including about 53,000 women:** Michael J. O'Dowd and Elliott E. Philipp, *The History of Obstetrics and Gynaecology* (New York: Parthenon Publishing Group, 1994), 99.

167 **Mt. Sinai Hospital Medical Center:** S. A. Myers and N. Gleicher, "A Successful Program to Lower Cesarean Section Rates," *New England Journal of Medicine* 319, no. 23 (1987): 1515.

Chapter Eleven

This chapter was based on interviews with many "freebirthers" and information gathered from Web sites and blogs. Laura Shanley's Web site is the most informative: www.unassistedchildbirth.com.

172 **They also share videos:** DVD of Pema Greenberg's birth, 4 July 2004; 13.33 minutes.

174 **Pat Carter:** Interviews with her daughters and with archival information collected from her daughters and Laura Shanley.

175 **she even got into *Look*:** "Natural Childbirth: This Mother of Nine Delivers Her Own Babies," *Look*, vol. 20, no. 23, 1956, 130–132.

178 **Female dolphins and elephants:** W. D. Thomas and D. Kaufman, *Dolphin Conferences, Elephant Midwives, and Other Astonishing Facts about Animals* (New York: Tarcher, Penguin, 1990), 179.

179 **Guatemalan study:** P. T. Ellison, *On Fertile Ground: A Natural History of Human Reproduction* (Cambridge: Harvard University Press, 2001), 55.

179 **monkey birth:** J. Lothian, "Do Not Disturb: The Importance of Privacy in Labor," *Journal of Perinatal Education* 13, no. 3 (2004): 4–6.

180 **narrow-pelvis-big-brain shape:** Personal interview with Wenda Trevathan, 17 July 2007; W. R. Trevathan, *Human Birth: An Evolutionary Perspective* (New York, Aldine De Gruyter, 1987).

181 **"It's hard for me to imagine":** Personal interview with Wenda Trevathan.

181 **Robbie Davis-Floyd's** *Childbirth and Authoritative Knowledge*: R. E. Davis-Floyd and C. F. Sargent, eds., *Childbirth and Authoritative Knowledge: Cross-Cultural Perspectives* (Berkeley: University of California Press, 1997), 83.

183 **Laurie Morgan:** L. A. Morgan, *The Power of Pleasurable Birth*. Lincoln, NE: Writers Club Press, 2003.

184 **Peter Ellison:** E-mail exchange with Peter Ellison 14 July 2007.

184 **Sarah Kilpatrick:** Telephone interview 5 July 2007.

185 **"What we hope to accomplish":** Personal interview with Joshua Copel.

185 **WHO study:** C. Guillermo, "WHO Systematic Review of Randomized Controlled Trials of Routine Antenatal Care," *Lancet* 357 (19 May 2001): 1565–1570.

Chapter Twelve

190 **"a craze unlike any that had come before":** B. H. Kevles, *Naked to the Bone: Medical Imaging in the Twentieth Century* (New York: Basic Books, 1997), 24, 25.

190 **By the 1930s, X-rays were a routine part:** L. M. Mitchell, *Baby's First Picture*. Toronto (University of Toronto Press, 2001).

191 **"double the risk of an early cancer death":** G. Greene, *The Woman Who Knew Too Much: Alice Stewart and the Secrets of Radiation* (Ann Arbor: University of Michigan Press, 2003), 83.

192 **Hence the name** *sonography*: P. A. Sullivan, "Commercial, Political, and Cultural Uses (and Abuses?) of Diagnostic Medical Ultrasound: Unforeseen Consequences, 1970–2005," PhD dissertation, Department of Preventive and Societal Medicine, University of Nebraska, Omaha, 2006, 109.

192 **"It is simply a question of refinement":** I. Donald, "On Launching a New Diagnostic Science," *American Journal of Obstetrics and Gynecology* 103, no. 5 (1969): 610.

192 **"I said this ring's a gestational sac":** I. Donald, "On Launching a New Diagnostic Science," *American Journal of Obstetrics and Gynecology* 103, no. 5 (1969): 610.

192 **"She goes to the loo and passes something":** A. Oakley, *The Captured Womb* (Oxford: Basil Blackwell, 1984), 163.

193 **"utmost vigilance must be maintained"**: I. Donald, "On Launching a New Diagnostic Science," *American Journal of Obstetrics and Gynecology* 103, no. 5 (1969): 628.

193 **"We must not forget that it took nearly half a century"**: I. Donald, "On Launching a New Diagnostic Science," *American Journal of Obstetrics and Gynecology* 103, no. 5 (1969): 624.

195 **"no longer capable of safely carrying birthing babies"**: R. B. Gold, "Ultrasound Imaging during Pregnancy," *Family Planning Perspectives* 16, no. 5 (1984): 242.

196 **Some mice were born with missing limbs:** Frigoletto interview.

197 **"loophole big enough to drive a truck through"**: P. A. Sullivan, "Commercial, Political, and Cultural Uses (and Abuses?) of Diagnostic Medical Ultrasound: Unforeseen Consequences, 1970–2005," PhD dissertation, Department of Preventive and Societal Medicine, University of Nebraska, Omaha, 2006, 109.

200 **"created false expectations that by having repeated"**: Natalie Angier, "Ultrasound and Fury: One Mother's Ordeal," *New York Times*, 26 November 1996; available at health.nytimes.com/health/guides/disease/clubfoot/news-and-features.html.

200 **"Ultrasound images have contributed"**: B. H. Kevles, *Naked to the Bone: Medical Imaging in the Twentieth Century* (New York: Basic Books, 1997), 246.

Chapter Thirteen

For a thorough review of the history of sperm banking, see C. R. Daniels and J. Golden, "Procreative Compounds: Popular Eugenics, Artificial Insemination and the Rise of the American Sperm Banking Industry," *Journal of Social History* 38, no. 1 (2004): 5–27.

205 **global sperm exports totaled somewhere between $50 and $100 million:** P. Zachary, "A Most Unlikely Industry Finds It Can't Resist Globalization's Call—Exporting Human Sperm Is a Fast-Growth Business, Banks in Denmark, U.S. Find," *Wall Street Journal*, 6 January 2000, 1; D. L. Spar, *The Baby Business: How Money, Science, and Politics Drive the Commerce of Conception* (Boston: Harvard Business School Press, 2006), 38.

205 **U.S. Food and Drug Administration:** C. R. Daniels and J. Golden, "Procreative Compounds: Popular Eugenics, Artificial Insemination and the

Rise of the American Sperm Banking Industry," *Journal of Social History* 38, no. 1 (2004): 5–27.

206 **Of the 20 . . . IVF but do not sell anonymous sperm:** D. Ridder, tissue bank interview, spokesperson for American Association of Tissue Banks, 2007.

207 **"This mechanical fertilization:** J. M. Sims, *Clinical Notes on Uterine Surgery with Special Reference to the Management of the Sterile Condition* (New York: William Wood & Co., 1867), 370.

208 **Silber's sperm facts:** R. H. Epstein, "The Male Mystery of All Life Came from the Sea," *Chicago Tribune*, 11 November 1992.

211 **F_{swim}/F_{shear}:** J, A. Riffell and R. K. Zimmer, "Sex and Flow: The Consequences of Fluid Shear for Sperm-Egg Interactions," *The Journal of Experimental Biology* 210 (2007): 3644–3660.

211 **"In terms of whether sperm swim straight or turn":** Riffel interview.

211 **Arnold of Villanova, a Spanish doctor:** F. N. L. Poynter, "Hunter, Spallanzani, and the History of Artificial Insemination," in *Medicine Science and Culture: Historical Essays in Honor of Oswei Temkin,* ed. L. G. Stevenson and R. P. Multhauf, 98 (Baltimore: The Johns Hopkins Press, 1968); O. Temkin, trans., *Soranus' Gynecology* (Baltimore: Johns Hopkins University Press, 1956).

212 **"artificial fructification.":** S. Harris, *Woman's Surgeon: The Life Story of J. Marion Sims* (New York: The MacMillan Co., 1950), 245.

212 **"Better let ancient families become extinct":** S. Harris, *Woman's Surgeon: The Life Story of J. Marion Sims* (New York: The MacMillan Co., 1950), 247.

214 **The first published account of donor sperm was in 1909:** C. R. Daniels and J. Golden, "Procreative Compounds: Popular Eugenics, Artificial Insemination and the Rise of the American Sperm Banking Industry," *Journal of Social History* 38, no. 1 (2004): 8.

214 **because "he has an intelligence quotient of 140":** F. I. Seymour, "Sterile Motile Spermatozoa Proved by Clinical Experimentation," *Journal of the American Medical Association* 112, no. 18 (1939): 1819.

215 **"artificial bastards":** C. R. Daniels and J. Golden, "Procreative Compounds: Popular Eugenics, Artificial Insemination and the Rise of the American Sperm Banking Industry," *Journal of Social History* 38, no. 1 (2004): 10.

215 **Illinois court that ruled donor sperm was adultery:** L. B. Andrews, *The Clone Age* (New York: Henry Holt, 1999), 86.

215 **What began as 10 banks:** C. R. Daniels and J. Golden, "Procreative Compounds: Popular Eugenics, Artificial Insemination and the Rise of the American Sperm Banking Industry," *Journal of Social History* 38, no. 1 (2004): 10.

216 **Cappy Rothman, description of home and sperm toys:** personal interview and visit to his home.

218 **sperminators:** L. B. Andrews, *The Clone Age* (New York: Henry Holt, 1999), 222–236.

218 **"greatest hindrance to medical advances":** C. M. Rothman, "Live Sperm, Dead Bodies," *Journal of Andrology* 20, no. 4 (July–August 1999): 457.

220 **California Cryobanks statistics:** visit and personal interviews with directors.

Chapter Fourteen

229 **"I know this sounds kind of stupid":** Interview with Lindsay Nohr, 2001.

231 **"the most amazing cell in the body":** The egg biology section was based on two articles by Sathananthan [A. H. Sathananthan, "Ultrastructure of the Human Egg," *Human Cell* 10, no. 1 (1997): 21–38; A. H. Sathananthan et al., "From Oogonia to Mature Oocytes: Inactivation of the Maternal Centrosome in Humans," *Microscopy Research and Technique* 69 (2006): 396–407], but primarily on several interviews with Dr. Teresa K. Woodruff, chief of the Division of Fertility Preservation at the Feinberg School of Medicine, Northwestern University, Chicago, and Jason Barritt, PhD, an embryologist and scientific director of Reproductive Medicine Associates in New York City.

240 **The American Society of Reproductive Medicine considers egg freezing experimental:** S. S. Wang, "Fertility Therapies under the Microscope," *Wall Street Journal*, 15 November 2007, 1.

241 **"Unlike its European counterparts, the U.S. government is notoriously loath to intervene":** D. L. Spar, *The Baby Business: How Money, Science, and Politics Drive the Commerce of Conception* (Boston: Harvard Business School Press, 2006), 228.

241 **"moved from an initial period of market anarchy":** D. L. Spar, *The Baby Business: How Money, Science, and Politics Drive the Commerce of Conception* (Boston: Harvard Business School Press, 2006), 279.

243 **All single women younger than 40:** Personal interview with C. Birrittieri, 17 December 2007.

244 **"What if law firms someday require female partners to freeze their eggs and wait even longer?":** L. Mundy, *Everything Conceivable: How Assisted Reproduction Is Changing Men, Women, and the World* (New York: Alfred A. Knopf, 2007), 329.

Bibliography

Introduction

"An Unusual but Harmless Occurrence." *Leeds Mercury*, 14 September 1889, 13.

Forbes, T. R. "The Social History of the Caul." *Yale Journal of Biology and Medicine* 53 (1952): 495–508.

Leavitt, J. W. *Brought to Bed: Childbearing in America, 1750–1950*. New York: Oxford University Press, 1986.

Chapter One: Eve's Doing: Birth From Antiquity Through the Middle Ages

Banks, A. C. *Birth Chairs, Midwives, and Medicine*. Jackson: University Press of Mississippi, 1999.

Caton, D. *What a Blessing She Had Chloroform*. New Haven: Yale University Press, 1999.

Crane, E. F., ed. *The Diary of Elizabeth Drinker: The Lifecycle of an Eighteenth Century Woman*. Boston: Northeastern University Press, 1994.

Dye, J. H. M. *Painless Childbirth or Healthy Mothers & Healthy Children*. Silver Creek, NY: The Local Printing House, 1884.

Ellison, P. T. *On Fertile Ground: A Natural History of Human Reproduction*. Cambridge: Harvard University Press, 2001.

Forbes, T. R. "The Social History of the Caul." *Yale Journal of Biology and Medicine* 53 (1952): 495–508.

Frieda, L. *Catherine de Medici Renaissance Queen of France.* New York: Fourth Estate, HarperCollins, 2003.

Hibbard, B. *The Obstetrician's Armamentarian.* San Anselmo, CA: Norman Publishing, 2000.

Hobby, E., ed. *The Midwives Book or the Whole Art of Midwifery Discovered. Women Writers in English 1350–1850.* New York: Oxford University Press, 1999.

Johnson, S. *The Ghost Map: The Story of London's Most Terrifying Epidemic--and How It Changed Science, Cities and the Modern World.* New York: Riverhead Books, 2006.

Lemay, Helen R. "Anthonius Guainerius and Medieval Gynecology." In *Women of the Medieval World,* edited by J. Kirshner and S. F. Wemple, 317–336. New York: Basil Blackwell, 1985.

Lemay, H. R. *Women's Secrets: A Translation of Pseudo-Albertus Magnus' De Secretis Mulierum with Commentaries.* Albany: State University of New York Press, 1992.

Lurie, S. "Euphemia Maclean, Agnes Sampson and Pain Relief during Labor in 16th Century Edinburgh." *Anaesthesia* 59, no. 8 (2004): 834–835.

Mauriceau, Francois. *The Diseases of Women with Child, and in the Child-Bed: as also the Best Means of Helping Them in Natural and Unnatural Labours, with Fit Remedies for the Several Indispositions of New-born Babes, to Which is Prefix'd an Exact Description of the Parts of Generation in Women,* 7th ed. Translated by H. T. Chamberlen. London: T. Cox and J. Clarke, 1763.

O'Dowd, Michael J., and Elliott E. Philipp. *The History of Obstetrics and Gynaecology.* New York: Parthenon Publishing Group, 1994.

Raju, T. N. K. "Soranus of Ephesus: Who Was He and What Did He Do?" *Historical Review and Recent Advances in Neonatal and Perinatal Medicine.* Mead Johnson Nutritional Division (1980).

Rosenberg, C. S., and Charles Rosenberg. "The Female Animal: Medical and Biological Views of Woman and Her Role in Nineteenth Century America." *The Journal of the American History* 60, no. 2 (1973): 332–356.

Rösslin, Eucharius. *When Midwifery Became the Male Physician's Province: The Sixteenth Century Handbook: The Rose Garden for Pregnant Women and Midwives.* Translated by Wendy Arons. Jefferson, NC: McFarland & Company, 1994.

Shorter, E. *A History of Women's Bodies.* New York: Penguin Books, 1982.

Temkin, O., trans.. *Soranus' Gynecology.* Baltimore: Johns Hopkins University Press, 1956.

Thomas, I., ed. *Culpeper's Book of Birth: A Seventeenth Century Guide to Having Lusty Children.* Devon, UK: Webb & Bower, 1985.

Thompson, L. *The Wandering Womb: A Cultural History of Outrageous Beliefs about Women.* New York: Prometheus Books, 1999.

Chapter Two: Men With Tools: Forceps Use From 1600s to 1880s

Bailey, P. E. "The Disappearing Art of Instrumental Delivery: Time to Reverse the Trend." *International Journal of Gynecology and Obstetrics* 91 (2005): 89–96.

Banks, A. C. *Birth Chairs, Midwives, and Medicine.* Jackson: University Press of Mississippi, 1999.

Das, K. *Obstetric Forceps: Its History and Evolution.* St. Louis: Mosby, 1929.

Dunn, P. M. "The Chamberlen Family (1560–1728) and Obstetric Forceps." *Archives of Disease Child Fetal Neonatal Edition* 81 (1999): F232–234.

Ehrenreich, B., and D. English. *Witches, Midwives and Nurses: A History of Women Healers.* New York: The Feminist Press at the City University of New York, 1973.

Engelmann, George. *Labor among Primitive Peoples: Sharing the Development of the Obstetric Science of To-Day.* St. Louis: Chambers, 1884.

Frieda, L. *Catherine de Medici.* New York: Fourth Estate, HarperCollins, 2005.

Hibbard, B. *The Obstetrician's Armamentarian.* San Anselmo, CA: Norman Publishing, 2000.

Holt, M. *The French Wars of Religion, 1562–1629.* New York: Cambridge University Press, 2005.

Mauriceau, Francois. *The Diseases of Women with Child, and in the Child-Bed: as also the Best Means of Helping Them in Natural and Unnatural Labours, with Fit Remedies for the Several Indispositions of New-born Babes, to Which is Prefix'd an Exact Description of the Parts of Generation in Women,* 7th ed. Translated by H. T. Chamberlen. London: T. Cox and J. Clarke, 1763.

Mitford, J. *The American Way of Birth.* New York: Plume, Williams Abrams, 1993.

Newell, F. S. "The Effect of Overcivilization on Maternity." *American Journal of the Medical Sciences* 136 (1908): 533–541.

O'Dowd, Michael J., and Elliott E. Philipp. *The History of Obstetrics and Gynaecology.* New York: Parthenon Publishing Group, 1994.

Parvin, T. "The Forceps." In *The Science and Art of Obstetrics.* Philadelphia: Lea and Brothers, 1895.

Patel, R., and D. Murphy. "Forceps Delivery in Modern Obstetrics Practice." *British Medical Journal* 328 (2004): 1302–1305.

Radcliffe, W. *Milestones in Midwifery and The Secret Instrument (The Birth of the Midwifery Forceps)*. San Francisco: Norman Publishing, 1989.

Scholten, C. M. *Childbearing in American Society: 1650–1850*. New York: New York University Press, 1985.

Shapira, Ian. "After Toil and Trouble, 'Witch' Is Cleared. Va. Resident's Quest Leads to Pardon for Woman Convicted in 1706." *Washington Post*, 12 July 2006, B01.

Wertz, R. W., and Dorothy C. Wertz. *Lying-In: A History of Childbirth in America*. New Haven: Yale University Press, 1989.

Chapter Three: Slave Women's Contribution to Gynecology

Applegate, D. *The Most Famous Man in America*. Houston: Three Leaves, 2007.

de Costa, C. M. "James Marion Sims: Some Speculations and a New Position." *Medical Journal of Australia* 178 (2003): 660–663.

Harris, S. *Woman's Surgeon: The Life Story of J. Marion Sims*. New York: The MacMillan Co., 1950.

Lerner, B. "Scholars Argue Over Legacy of Surgeon Who Was Lionized, Then Vilified." *New York Times*, 28 October 2003.

Loudon, I. *Death in Childbirth: An International Study of Maternal Care and Maternal Mortality 1800–1950*. Oxford: Clarendon Press, 1992.

Marr, J. P. *James Marion Sims: The Founder of the Woman's Hospital in the State of New York*. New York: The Woman's Hospital, 1949.

McGregor, D. K. *From Midwives to Medicine: The Birth of American Gynecology*. New Brunswick: Rutgers University Press, 1988.

Rice, J. H., Jr. Eulogy. *James Marion Sims Memorial*. Columbia, SC: South Carolina Medical Association, 1929.

Rosenberg, C. "Belief and Ritual in Antebellum Medical Therapeutics." In *Major Problems in the History of American Medicine and Public Health*, edited by J. H. Warner and J. A. Tighe. New York: Houghton Mifflin, 2001.

Savitt, T. L. *Medicine and Slavery: The Diseases and Health Care of Blacks in Antebellum Virginia*. Chicago: University of Illinois Press, 1978.

Scholten, C. M. *Childbearing in American Society: 1650–1850*. New York: New York University Press, 1985.

Sims, J. M. *The Story of My Life*. New York: D. Appleton and Co., 1898. Reprint, Whitefish, MT: Kessinger's Publishing Rare Reprints, 2007.

Wall, L. L. "The Medical Ethics of Dr. J. Marion Sims: A Fresh Look at the Historical Record." *Journal of Medical Ethics* 32, no. 6 (2006): 346–350.

Wall, L. L. "Obstetric Vesicovaginal Fistula as an International Public-Health Problem." *Lancet* 368, no. 9542 (2006): 1201–1209.

Warner, John H., and Janet A. Tighe, eds. *Major Problems in the History of American Medicine and Public Health*, chaps. 4–6. Boston: Houghton Mifflin, 2001.

Washington, H. *Medical Apartheid: The Dark History of Medical Experimentation on Black Americans from Colonial Times to the Present*. New York: Doubleday, 2006.

Chapter Four: Dying to Give Birth: Maternal Mortality into the Twentieth Century

Baker, J. S. "Maternal Mortality in the United States." *Journal of the American Medical Association* 89, no. 24 (1927): 2016–2017.

Barry, J. M. *The Great Influenza: The Story of the Deadliest Pandemic in History*. New York: Penguin Books, 2004.

Bromley, D. D. "What Risk Motherhood." *Harper's Magazine*, June 1929, 11–22.

Carter, K. C. *Ignaz Semmelweis: The Etiology, Concept, and Prophylaxis of Childbed Fever*. Madison: University of Wisconsin Press, 1983.

Carter, K. C. *Childbed Fever: A Scientific Biography of Ignaz Semmelweis*. New Brunswick, NJ: Transaction Publishers, 2005.

Gelis, J. *History of Childbirth: Fertility, Pregnancy, and Birth in Early Modern Europe*. Boston: Northeastern University Press, 1991.

Hoosen, B. V. *Petticoat Surgeon*. Chicago: People's Book Club, 1947.

Irving, F. C. *Safe Deliverance*. Boston: Houghton Mifflin, 1942.

Lambert, S., and H. Painter. "Fever in the Puerperal Woman." *Dispensary Report*, 1893. Scrapbooks of the Lying-In Hospital: 1890–1937. Medical Center Archives of New York-Presbyterian/Weill Cornell, New York, NY.

"Lane Bryant Malsin: Fashion Revolutionary." In *Blessings of Freedom: Chapters in American Jewish History*, edited by Michael Feldberg. Hoboken, NJ: KTAV, 2002. Originally published by the American Jewish Historical Society.

Loudon, I. *Death in Childbirth: An International Study of Maternal Care and Maternal Mortality 1800–1950*. Oxford: Clarendon Press, 1992.

Martin, E. "Puerperal Fever: As It Occurred in the General Hospital of Vienna during the First Six Months of 1834." *Lancet* 26, no. 675 (1836): 649–652.

Marx, S. "The Bacteriology of the Puerperal Uterus: Its Relation to the Treatment of the Parturient State." *American Journal of Obstetrics* 48, no. 3 (1903): 301–323.

Nuland, S. *The Doctors' Plague: Germs, Childbed Fever, and the Strange Story of Ignac Semmelweis.* New York: W.W. Norton, 2003.

Pancoast, S. M. *The Ladies' New Medical Guide: An Instructor, Counsellor, and Friend in All the Delicate and Wonderful Matters Peculiar to Women.* Philadelphia: World Bible House, 1890.

Schlereth, T. J. "Living and Dying." In *Victorian America: Transformations in Everyday Life*, chap. 8. New York: HarperPerennial, 1992.

Scrapbooks of the Lying-In Hospital: 1890–1937. Medical Center Archives of New York-Presbyterian/Weill Cornell, New York, NY. The scrapbooks contain articles from many local papers, including the *Brooklyn Eagle*, which were cited in the chapter.

Shoemaker, G. "Transactions of the Section of Gynecology of the College of Physicians of Philadelphia." *American Journal of Obstetrics* 45 (1902): 563–566.

Stockman, A. B. *Tokology.* New York: R. F. Fenno & Company, 1911.

Tilton, E. M. *Amiable Autocrat.* New York: Henry Schuman, 1947.

Vineberg, H. "The Treatment of Puerperal Sepsis." *American Journal of Obstetrics* 48 (September 1903).

Waldorf Astoria Invitation, Scrapbooks of the Lying-In Hospital: 1890–1937. Medical Center Archives of New York-Presbyterian/Weill Cornell, New York, NY.

Watson, B. P. "What Can We Do to Improve Our Present Puerperal Mortality Rate?" *America Journal of Obstetrics and Gynecology* 19 (1930): 433–440.

"What Lydia E. Pinkham's Vegetable Compound Is Doing for Women." Advertisement in the *Boston Daily Globe*, 1 April 1900, 9.

Worboys, M. *Spreading Germs: Disease Theories and Medical Practice in Britain 1865–1900.* New York: Cambridge University Press, 2000.

Zeitz, J. *Flapper: A Madcap Story of Sex, Style, Celebrity and the Women Who Made America Modern.* New York: Three Rivers Press, 2006.

Chapter Five: Leaving Home: New York's Lying-In and the Growth of Maternity Wards

"Birth of a Baby," *Life*, 11 April 1938, 32–37.

Cody, L. F. "Living and Dying in Georgian London's Lying-In Hospitals." *Bulletin of the History of Medicine* 78, no. 2 (summer 2004): 304–348.

DeLee, J. B. *The Principles and Practice of Obstetrics.* Philadelphia: W. B. Saunders, 1920.

Harrar, J. A. *The Story of the Lying-In Hospital of the City of New York*. New York: The Society of the Lying-In, 1938.

Hoosen, B. V. *Petticoat Surgeon*. Chicago: People's Book Club, 1947.

Irving, F. C. *Safe Deliverance*. Boston: Houghton Mifflin, 1942.

Leavitt, J. W. *Brought to Bed: Childbearing in America, 1750–1950*. New York: Oxford University Press, 1986.

Lambert, S., and H. Painter. "Fever in the Puerperal Woman." *Dispensary Report*, 1893. Scrapbooks of the Lying-In Hospital: 1890–1937. Medical Center Archives of New York-Presbyterian/Weill Cornell, New York, NY.

Radcliffe, W. *Milestones in Midwifery and The Secret Instrument (The Birth of the Midwifery Forceps)*. San Francisco: Norman Publishing, 1989.

Rosenberg, C. E. *The Care of Strangers: The Rise of America's Hospital System*. New York: Basic Books, 1987.

Rosner, D. *A Once Charitable Enterprise: Hospitals and Health Care in Brooklyn and New York 1885–1915*. New York: Cambridge University Press, 1982.

Scrapbooks of the Lying-In Hospital 1890–1937. Medical Center Archives of New York-Presbyterian/Weill Cornell, New York, NY.

Wertz, R. W., and Dorothy C. Wertz. *Lying-In: A History of Childbirth in America*. New Haven: Yale University Press, 1989.

Chapter Six: Birth Is but a Sleep and Forgetting

Many of the clippings from newspapers are from the Eliza Taylor Ransom Papers, M-61, 12 Volume 2, Scrapbook on Twilight Sleep 1914–1915. Schlesinger Library, Radcliffe Institute for Advanced Study, Harvard University.

"Anti–Twilight Sleep Association to Fight Twilight Sleep: A Brooklyn Woman to Start Association to Oppose the Treatment." *New York Times*, 13 August 1915, 5.

Bromley, D. D. "Lifting the Curse of Eve." *Woman's Journal* 1 (October 1927): 8–10, 36–37.

Caton, D. *What a Blessing She Had Chloroform*. New Haven: Yale University Press, 1999.

Collins, G. *America's Women: Four Hundred Years of Dolls, Drudges, Helpmates, and Heroines*. New York: William Morrow, 2003.

Coontz, S. *Marriage, A History*. New York: Viking, 2005.

Coontz, S. *The Way We Never Were: American Families and the Nostalgia Trap*. New York: Basic Books, 2000.

"Drug Boon to Women: New Treatment for Childbirth Called Medical Mercy." *Washington Post*, 27 August 1914, 4.

Feldberg, M., ed. *Blessings of Freedom: Chapters in American Jewish History*. Hoboken, NJ: KTAV, 2002. Also available at www.ajhs.org.

Hamilton, J. "The Twilight Sleep: Why Physicians Are Conservative and Why Women May Not Expect Too Much from It." *The Evening Sun*, 23 April 1914.

Harris, Hazel. *New York Times*, 28 May 1914.

Hutchinson, W. "The Kind of Woman Who Ought to Have Babies." *Washington Post*, 16 July 1916, MT8.

Kaplan, J. *When the Astors Owned New York*. New York: Plume, 2006.

"Lane Bryant Malsin: Fashion Revolutionary." In *Blessings of Freedom: Chapters in American Jewish History*, edited by Michael Feldberg. Hoboken, NJ: KTAV, 2002. Originally published by the American Jewish Historical Society.

Leavitt, J. W. "Birthing and Anesthesia: The Debate Over Twilight Sleep." *Signs: Journal of Women in Culture and Society* 6, no.1 (1980): 147–164.

Leavitt, J. W. *Brought to Bed: Childbearing in America 1750–1950*. New York: Oxford University Press, 1986.

Leupp, C., and B. J. Henrick. "Twilight Sleep in America." *McClure's Magazine*, April 1915, 25–36.

"Mrs. Dennett Guilty in Sex Booklet Case." *New York Times*, 29 April 1929, 31.

Pancoast, S. M. *The Ladies' New Medical Guide: An Instructor, Counsellor, and Friend in All the Delicate and Wonderful Matters Peculiar to Women*. Philadelphia: World Bible House, 1890.

Polak, J. O. "A Study of Twilight Sleep with a Critical Analysis of the Cases at the Long Island College Hospital." *New York Medical Journal* (February 13, 1915).

Richardson, A. S. "Safety First for Mother." Newspaper article, 1 May 1915, 24. From the Eliza Taylor Ransom Papers.

Richardson, A. S. "What Every Mother Wants to Know about Her Baby." *Atlantic Constitution*, 4 August 1914, 12.

Sanghavi, D. "The Mother Lode of Pain; Some Women Insist on a Drug-Free Childbirth—Even Though It Might Be Agonizing—While Others Opt for a Numbing Epidural. Is This All Part of the Simmering Debate Between Natural and Modern Medicine, or Are Some Women Embracing Labor Pain for a More Heroic Cause?" *Boston Globe Magazine*, 23 July 2006, 18–21, 28–29.

"Says Twilight Sleep Is Not Dead." *Washington Post*, 29 August 1915, A8.

Stockman, A. B. *Tokology*. New York: R. F. Fenno & Company, 1911.

Thompson, V. "Motherhood without Fear." Newpaper article. From the Eliza Taylor Ransom Papers, microfilm M-6, 12 vol. 2, Schlesinger Library, Radcliffe Institute for Advanced Studies, Harvard University.

Tracy, M., and M. Boyd. *Painless Childbirth: A General Survey of All the Painless Methods, with Special Stress on "Twilight Sleep" and Its Extension to America*. New York: Frederick A. Stokes Company, 1915.

Tracy, M., and C. Leupp. "Painless Childbirth." *McClure's Magazine*, vol. 43, 1914.

"Twilight Sleep Is Necessity, Not Luxury." *International News Service*, 31 May 1916.

Wolf, Jacqueline H. *Deliver Me from Pain: Anesthesia and Birth in America*. Baltimore: Johns Hopkins University Press, 2009.

Wolff, E. "The Bridge of Dreams." *New York Times*, 28 May 1914, 12.

Zeitz, J. *Flapper: A Madcap Story of Sex, Style, Celebrity and the Women Who Made America Modern*. New York: Three Rivers Press, 2006.

Chapter Seven: What Was She Thinking? Freud Meets Fertility

Many of the scientific publications, including Jacobson's article, as well as notes from psychiatrists and correspondence, were reprinted with permission from the extensive collection of Dr. Viola W. Bernard included in Boxes 57 and 58, the Archives and Special Collections, A. C. Long Health Sciences Library, Columbia University.

Benedek, T. "Infertility as a Psychosomatic Defense." *Fertility and Sterility* 3, no. 6 (1952): 527–537.

Berga, S. "Psychiatry and Reproductive Medicine." In *Kaplan & Sadock's Comprehensive Textbook of Psychiatry*, 7th ed., vol. 2, edited by Benjamin Sadock and Virginia Sadock, 1935–1952. Philadelphia: Lippincott Williams & Wilkins, 2000.

Berga, S. "Recovery of Ovarian Activity in Women with Functional Hypothalamic Anmenorrhea Who Were Treated with Cognitive Behavior Therapy." *Fertility and Sterility* 80, no. 4 (2003): 976–980.

Benedek, T. "Infertility as a Psychosomatic Defense." *Fertility and Sterility* 3, no. 6 (1952): 527–537.

Benedek, Therese, and Boris B. Rubenstein. "The Correlations Between Ovarian Activity and Psychodynamic Processes: II. The Menstrual Phase." *Psychosomatic Medicine* 1, no. 4 (October 1939): 461–485.

Benedek, Therese, George Ham, et al. "Some Emotional Factors in Infertility." *Psychosomatic Medicine* 15, no. 5 (1953): 485–498. From the Viola W. Bernard Archives.

Bernard, V. "Abstract of Talk at the National Conference of Social Work for Planned Parenthood on Psychiatric Aspects of Infertility in Women," April 15, 1947. From Box 58, the Archives and Special Collections, A. C. Long Health Sciences Library, Columbia University.

Castle, Molly. "New Hope for the Sterile." *Pageant*, December 1944, 62–65.

Coontz, S. *The Way We Never Were: American Families and the Nostalgia Trap.* New York: Basic Books, 2000.

Dershimer, Frederick W. "Influence of Mental Attitudes in Childbearing." *American Journal or Obstetrics and Gynecology* 31, no. 3 (March 1936): 444.

Domar, Alice. *Conquering Infertility: Dr. Alice Domar's Mind/Body Guide to Enhancing Fertility and Coping With Infertility.* New York: Penguin, 2004.

"Fertility Fantasies." *Time*, 7 June 1948, 80.

Friess, C. "What We Know About Miscarriage." *Woman's Home Companion* 82 (1955): 4–6.

Jacobson, E. "A Case of Sterility." *Psychoanalytic Quarterly* 15, no. 3 (1946): 330–350. From the Viola W. Bernard Archives.

Jones, Ernst. "Psychology and Childbirth." *Lancet* 242 (6 June 1942): 695–696.

"Just a Bundle of Nerves." *Hygiea* (April 1940).

Kelley, Kenneth. "Sterility in the Female." *Psychosomatic Medicine* 4, no.2 (1942).

Kenyon, J. H. "Tired Mother," *Good Housekeeping* 110, January 1940, 685–688.

King, James E. "Presidential Address: American Association of Obstetricians, Gynecologists, and Abdominal Surgeons," *American Journal of Obstetrics and Gynecology* 39, no. 2 (February 1939): 180–181.

Lundberg, F., and M. Farnham. *Modern Woman: The Lost Sex.* New York: Harper & Brothers, 1947.

Marsh, E. M., and A. M. Vollmer. "Possible Psychogenic Aspects of Fertility." *Fertility and Sterility* 1 (1951): 70–79.

McLaughlin, L. *The Pill, John Rock, and the Church.* Boston: Little, Brown and Company, 1982.

Meninger, K. A. "Somatic Correlations with Unconscious Repudiation of Fertility in Women." *Journal of Nervous and Mental Diseases* 89 (April 1939): 514–527. Also in *Bulletin of the Meninger Clinic* 3 (July 1939): 106–121.

Miller, J. M. "Psychogenic Menorrhagia." *Medical Journal and Record* (15 July 1931): 84–86.

Miller, L. M. "Changing Life Sensibly: New Hormone Treatments for Menopause." *Independent Woman*, September 18, 1939, 297. Also in *Reader's Digest* vol. 35, October 1939, 101–103.

Miller, M. C. "Facts about Menopause." *Hygiea* (August 1940): 692–694.

Mintz, S., and S. Kellogg. *Domestic Revolutions: A Social History of American Family Life*. New York: The Free Press, 1988.

O'Dowd, Michael J., and Elliott E. Philipp. *The History of Obstetrics and Gynaecology*. New York: Parthenon Publishing Group, 1994.

Organotherapy in General Practice. New York, G. W. Carnick Co., 1924.

Orr, D. W. "Pregnancy Following Decision to Adopt." *Psychosomatic Medicine* 3, no. 4 (1941): 441–446.

Oudshoorn, N. *Beyond the Natural Body: An Archeology of Sex Hormones*. New York: Routledge, 1994.

Porter, R. *The Greatest Benefit to Mankind*. New York: W. W. Norton, 1997.

Pruette, L. "Critical Days for Mother." *Parents* 35, 15 May 1940.

Resolve. Information pamphlet of the National Infertility Association, 1974. From the Viola W. Bernard Archives.

Rosenberg, C. S., and Charles Rosenberg. "The Female Animal: Medical and Biological Views of Woman and Her Role in Nineteenth Century America." *The Journal of the American History* 60, no. 2 (1973): 332–356.

Rothman, S., and D. Rothman. *The Pursuit of Perfection: The Promise and Perils of Medical Enhancement*. New York: Pantheon, 2003.

Rubin, I.C. "Diagnostic Procedures in the Investigation of Sterility in the Female: Evalution of Their Clinical Importance." In *Collected Papers of Dr. I. C. Rubin 1910–1954*. From the Viola W. Bernard Archives.

Rubin, I. C. "Ovarian Hypofunction, Habitually Delayed and Scanty Menstruation in Relation to Sterility and Lowered Fertility." Presented at the 54th annual meeting of the American gynecological society, Old Point Comfort, VA, May 20–22,1929.

Rubin, I. C. "Thirty Years of Progress in Treating Infertility." *Fertility and Sterility* 1 (September 1950): 5, 389–406.

Siegler, S. L. (1948). "Why Sterility." *The Urologic and Cutaneous Review* 52, no. 10 (1948): 571–575.

"Sterility and Neurotics." *Time*, 16 June 1952, 81.

Weiss, E. "Psychosomatic Problems in Fertility." *Human Fertility* 10, no. 3 (1945): 74–78.

Chapter Eight: It's Only Natural

Most of the articles in this chapter are from the Herbert Thoms Scrapbook of the Herbert Thoms Collection, Historical Library, Cushing/Whitney Medical Library, Yale University.

"Baby Born and No Pain at All: Author of Childbirth without Fear Dr. Grantly Read Visiting America." *The Herald* (Bridgeport), 19 January 1947. From the Herbert Thoms Collection.

"CDC on Infant and Maternal Mortality in the United States: 1900–99." *Population and Development Review* 25, no. 4 (1999): 821–826.

Dick-Read, G. *The Natural Childbirth Primer.* New York: Harper & Brothers, 1955.

Family Centered Maternity and Infant Care: Report of the Committee on Rooming-In of the Josiah Macy Jr. Foundation Conference on Problems in Infancy and Early Childhood. New York: Josiah Macy Jr. Foundation, 1950.

Freedman, L., and V. M. Ferguson. "The Question of 'Painless Childbirth' in Primitive Cultures." *The American Journal of OrthoPsychiatry* 20, no. 2 (1950).

G.A. Letter to the Editor. *West Hartford News*, 1950. From the Herbert Thoms Collection.

Galton, L. "Motherhood without Misery." *Collier's*, 1946. From the Herbert Thoms Collection.

Goodrich, Frederick W., Jr. "The Theory and Practice of Natural Childbirth." *Yale Journal of Biology and Medicine* 25, no. 6 (June 1953): 529–534.

Goodrich, F., Jr., and Thoms, Herbert (1948). "A Clinical Study of Natural Childbirth: A Preliminary Report from a Teaching Ward Service." *American Journal of Obstetrics and Gynecology* 56, no. 5 (1948): 875–883.

Jackson, E. B. "Mothers and Babies Together." *Parents Magazine*, vol. 22, 1947, 18–19, 146–149.

Jackson, E. B. "Rooming In Gives Baby a Good Start." *The Child* 12 (1948): 162–165.

Jackson, E. B. "The Rooming-In Plan for Mothers and Infants." *American Journal of Obstetrics and Gynecology* 65 (1948): 707–711.

Kartchner, F. "A Study of the Emotional Reactions During Labor." *American Journal of Obstetrics and Gynecology* 60, no. 1 (1950): 19–29.

Kemp, C. "British Doctor Claims Suffering Results from Physical Derangement or Misunderstanding on Part of the Mother." *Courier Journal* (Louisville, KY), 23 February 1947. From the Herbert Thoms Collection.

Kroger, William. *Childbirth with Hypnosis*. New York: Doubleday, 1961.

Lindsay, Malvina. "Modern Trek to Nature," *Washington Post*, 27 January 1949, 10.

Loudon, I. *Death in Childbirth: An International Study of Maternal Care and Maternal Mortality 1800–1950*. Oxford: Clarendon Press, 1992.

Mandy, A. J., et. al. "Is Natural Childbirth Natural." *Psychosomatic Medicine* 14, no. 6 (1952): 431–438.

O'Dowd, Michael J., and Elliott E. Philipp. *The History of Obstetrics and Gynaecology*. New York: Parthenon Publishing Group, 1994.

Ratcliff, J. D. "Miscarriage." *Woman's Home Companion*, vol. 77, 1950, 42.

Schultz, G. "Cruelty on the Maternity Wards." *Ladies Home Journal*, May 1958, 44–45, 152, 155.

Studdiford, William. "Childbirth Difference Is Tale of Two Cities." *West Hartford News* (CT), 19 January 1950.

Thomas, M. *Post-War Mothers: Childbirth Letters to Dr. Grantly Dick-Read 1946–1956*. Rochester, NY: University of Rochester Press, 1997.

Wertz, R. W., and Dorothy C. Wertz. *Lying-In: A History of Childbirth in America*. New Haven: Yale University Press, 1989.

Wolf, Jacqueline H. *Deliver Me from Pain: Anesthesia and Birth in America*. Baltimore: The Johns Hopkins University Press, 2009.

Chapter Nine: Toxic Advice and a Deadly Drug: DES

The information about the Smiths and the relevant correspondence are from DES-related records in the George and Olive Watkins Smith Papers [H MS c68] at the Center for the History of Medicine, Countway Library of Medicine, an Alliance of the Boston Medical Library and the Harvard Medical School.

Alvarez, W. C. "New Light on the Mechanisms by which Nervousness Causes Discomfort." *Journal of the American Medical Association* 115, no. 12 (1939): 1010–1013.

Apfel, R. J. "Psychoanalysis and Infertility." *International Journal of Psychoanalysis* 83 (2002): 85–104.

Apfel, R. and S. M. Fisher. *To Do No Harm*. New Haven: Yale University Press, 1984.

Asbell, B. *The Pill: A Biography of the Drug That Changed the World*. New York: Random House, 1995.

Bell, S. E. "Gendered Medical Science: Producing a Drug for Women." *Feminist Studies* 21, no. 3 (1995): 469–500.

Bell, S. E. "The Synthetic Compound Diethylstilbestrol (DES) 1938–1941: The Social Construction of a Medical Treatment." PhD dissertation, Department of Sociology, Brandeis University, Waltham, MA, 1980, 376.

Bichler, J. *DES Daughter.* New York: Avon, 1981.

Bryant, D. "The DES Odyssey of Pat Cody." *San Francisco Sunday Examiner and Chronicle,* 18 March 1979, 17–22.

Caton, D. *What a Blessing She Had Chloroform: The Medical and Social Response to the Pain of Childbearing from 1800 to Present.* New Haven, CT: Yale University Press, 1999.

Cody, P. "The Silent Epidemic." Berkeley, California, 2006, 185.

Dieckmann, W. J., et al. "Does the Administration of Diethylstilbestrol during Pregnancy Have Therapeutic Value?" *American Journal of Obstetrics and Gynecology* 66, no. 5 (1953): 1062–1081.

Fishbein, M. "Too Many Miscarriages." *Woman's Home Companion,* vol. 71, 1944, 20, 37.

Friess, C. "What We Know about Miscarriage." *Woman's Home Companion,* vol. 82, 1955, 4–6.

Gladwell, M. "John Rock's Error." *New Yorker,* 13 March 2000, 52–63.

Herbst, A. L., Howard Ulfelder, and David Poskanzer. "Adenocarcinoma of the Vagina: Association of Maternal Stilbestrol Therapy with Tumor Appearance in Young Women." *The New England Journal of Medicine* 284, no.16 (1971): 878–881.

"Just a Bundle of Nerves." Advertisement in *Hygeia* (1940): 293.

Leech, P. N. "Preliminary Report of the Council: Stilbestrol." *Journal of the American Medical Association* 113, no. 26 (1939): 2312.

Lovegrove, T. Letter to Olive Smith, from Wonghan Hills, West Australia, 25 November 1965.

McLaughlin, L. *The Pill, John Rock, and the Church.* Boston: Little, Brown and Company, 1982.

Orenberg, C. L. *DES: The Complete Story.* New York: St. Martin's Press, 1981.

Pitkin, R. M. "Classic Article: Herbst, A. L., et al., Vaginal and Cervical Abnormalities after Exposure to Stilbesterol in Utero." *Obstetrics and Gynecology* 102, no. 2 (2003): 222. [Herbst et. al. article originally appeared in *Obstetrics and Gynecology* 40(1972): 287–298.]

Quindlen, Anna. *New York Times,* 9 May 1993, section 4, 15.

Ratcliff, J. D. "Miscarriage." *Woman's Home Companion,* vol. 77, 1950, 42.

Seaman, B. *The Greatest Experiment Ever Performed on Women.* New York: Hyperion, 2003.

Shorr, E., Frank Robinson, and George N. Papanicolaou. "A Clinical Study of the Synthetic Estrogen Stilbestrol." *Journal of the American Medical Association* 113, no. 26 (1939): 2312–2318.

Stephens, T. D., and Rock Brynner. *Dark Remedy: The Impact of Thalidomide and Its Revival as a Vital Medicine.* New York: Perseus Publishing, 2001.

Weinstein, J. B. Memorandum and order in *Braune v. Abbott Laboratories, et. al.,* Eastern District of New York, United States District Court, 1995.

Chapter Ten: From Kitchen-Table Surgery to the Art of the C-Section

Churchill, H. *Caesarean Birth: Experience, Practice and History.* Cheshire, England: Books for Midwives Press, an imprint of Elsevier Limited, 1997.

Ecker, J., and F. Frigoletto Jr. "Cesarean Delivery and the Risk-Benefit Calculus." *New England Journal of Medicine* 356, no. 9 (2007): 885–888.

Frank, E., John Rock, and Danielle Sara. "Characteristics of Female Obstetricians-Gynecologists in the United States." *Obstetrics and Gynecology* 94, no. 5 (1999): 659–665.

Greene, Michael F. "Vaginal Birth after Cesarean Revisited," *New England Journal of Medicine* 351, no. 25 (2004): 2647–2649.

Hooker, R. S. *Maternal Mortality in New York City: A Study of All Puerperal Deaths 1930–1932.* New York: The Commonwealth Fund, 1933.

Landon, Mark B., John C. Hauth, Kenneth J. Leveno, Catherine Y. Spong, et al. "Maternal and Perinatal Outcomes Associated with a Trial of Labor after Prior Cesarean Delivery." *The New England Journal of Medicine* 351, no. 25 (2004): 2581–2589.

Leavitt, J. W. *Brought to Bed: Childbearing in America, 1750–1950.* New York: Oxford University Press, 1986.

Menacker, F. "Trends in Cesarean Rates for First Births and Repeat Cesarean Rates for Low-Risk Women: United States, 1990–2003." *National Vital Studies Reports* 54, no. 4 (2005): 1–12. Available at www.cdc.gov/nchs.

Myers, S. A., and N. Gleicher. "A Successful Program to Lower Cesarean Section Rates." *New England Journal of Medicine* 319, no. 23 (1987): 1511–1516.

Newell, F. S. "The Effect of Overcivilization on Maternity." *American Journal of the Medical Sciences* 136 (1908): 533–541.

O'Dowd, Michael J., and Elliott E. Philipp. *The History of Obstetrics and Gynaecology.* New York: Parthenon Publishing Group, 1994.

Savitt, T. L. *Medicine and Slavery: The Diseases and Health Care of Blacks in Antebellum Virginia.* Chicago: University of Illinois Press, 1978.

Shorter, E. *A History of Women's Bodies.* New York: Penguin Books, 1982.

Speert, H. *Obstetrics and Gynecology in America.* Baltimore: Waverly Press, 1980.

Wertz, R. W., and Dorothy C. Wertz. *Lying-In: A History of Childbirth in America.* New Haven: Yale University Press, 1989.

Worboys, M. *Spreading Germs: Disease Theories and Medical Practice in Britain, 1865–1900.* New York: Cambridge University Press, 2000.

Chapter Eleven: Freebirthers

Buckley, S. J. M. *Gentle Birth, Gentle Mothering.* Anstead, Australia: One Moon Press, 2009.

Carter, P. "Come Gently, Sweet Lucinda." Titusville, FL. From the archives of Laura Shanley, Boulder, Colorado, 1957.

Davis-Floyd, R. E., and C. F. Sargent, eds. *Childbirth and Authoritative Knowledge: Cross-Cultural Perspectives.* Berkeley: University of California Press, 1997.

Ellison, P. T. *On Fertile Ground: A Natural History of Human Reproduction.* Cambridge: Harvard University Press, 2001.

Freeze, R. A. S. "Born Free: Unassisted Childbirth in North America." *American Studies.* University of Iowa. PhD (2008): 355.

Gelis, J. *History of Childbirth: Fertility, Pregnancy, and Birth in Early Modern Europe.* Boston: Northeastern University Press, 1991.

Greenberg. DVD of Pema Greenberg's birth, 4 July 2004. 13.33 minutes.

Guillermo, C. "WHO Systematic Review of Randomized Controlled Trials of Routine Antenatal Care." *Lancet* 357 (19 May 2001): 1565–1570

Lothian, J. "Do Not Disturb: The Importance of Privacy in Labor." *Journal of Perinatal Education* 13, no. 3 (2004): 4–6.

Morgan, L. A. *The Power of Pleasurable Birth.* Lincoln, NE: Writers Club Press, 2003.

"Natural Childbirth: This Mother of Nine Delivers Her Own Babies." *Look* 20, no. 23 (1956): 130–132.

Seaman, J. *A Clear Road to Birth.* HomeBirthVideos.com, 2000. 56 minutes.

Shulz, G. D. "The Uninsulted Child." *Ladies Home Journal,* June 1956, 60–63, 135–136.

Thomas, W. D., and D. Kaufman. *Dolphin Conferences, Elephant Midwives, and Other Astonishing Facts about Animals.* New York: Tarcher (Penguin), 1990.

"Titusville Woman Has 7th Child without Aid." Jacksonville, FL, *Times Union*, 21 August 1956.

Trevathan, W. R. *Human Birth: An Evolutionary Perspective*. New York, Aldine De Gruyter, 1987.

Chapter Twelve: Womb with a View

Angier, Natalie. "Ultrasound and Fury: One Mother's Ordeal," *New York Times*, 26 November 1996. health.nytimes.com/health/guides/disease/clubfoot/news-and-features.html.

Donald, I. "On Launching a New Diagnostic Science." *American Journal of Obstetrics and Gynecology* 103, no. 5 (1969): 609–628.

Gold, R. B. "Ultrasound Imaging During Pregnancy." *Family Planning Perspectives* 16, no. 5 (1984): 240–243.

Goldberg, B. "Obstetric US Imaging: The Past 40 Years." *Radiology* 215, no. 3 (2000): 622–639.

Greene, G. *The Woman Who Knew Too Much: Alice Stewart and the Secrets of Radiation*. Ann Arbor: University of Michigan Press, 2003.

Kevles, B. H. *Naked to the Bone: Medical Imaging in the Twentieth Century*. New York: Basic Books, 1997.

Mitchell, L. M. *Baby's First Picture*. Toronto: University of Toronto Press, 2001.

Oakley, A. *The Captured Womb*. Oxford: Basil Blackwell, 1984.

Sullivan, P. A. "Commercial, Political, and Cultural Uses (and Abuses?) of Diagnostic Medical Ultrasound: Unforeseen Consequences, 1970–2005." PhD dissertation, Department of Preventive and Societal Medicine, University of Nebraska, Omaha, 2006, 109.

Chapter Thirteen: Sperm Shopping

Andrews, L. B. *The Clone Age: Adventures in the New World of Reproductive Technologies*. New York: Holt, 1999.

Daniels, C. R., and J. Golden. "Procreative Compounds: Popular Eugenics, Artificial Insemination and the Rise of the American Sperm Banking Industry." *Journal of Social History* 38, no. 1 (2004): 5–27.

Epstein, R. H. "The Male Mystery of All Life Came from the Sea." *Chicago Tribune*, 1992.

Harris, S. *Woman's Surgeon: The Life Story of J. Marion Sims*. New York: The MacMillan Co., 1950.

Hogue, F. "Social Eugenics." *The Los Angeles Times*, 1939, 122.

Patrizio, P., Anna C. Mastroianni, and Luigi Mastroianni. "Disclosure to Children Conceived with Donor Gametes Should Be Optional." *Human Reproduction* 16, no. 10 (2001): 2036–2038.

Plotz, D. *The Genius Factory: The Curious History of the Nobel Prize Sperm Bank*. New York: Random House, 2005.

Poynter, F. N. L. "Hunter, Spallanzani, and the History of Artificial Insemination." In *Medicine Science and Culture: Historical Essays in Honor of Oswei Temkin*, edited by L. G. Stevenson and R. P. Multhauf, 97–113. Baltimore: The Johns Hopkins Press, 1968.

Ridder, D. Tissue Bank interview, spokesperson for American Association of Tissue Banks, 2007.

Riffell, J. A., and R. K. Zimmer. "Sex and Flow: The Consequences of Fluid Shear for Sperm-Egg Interactions." *The Journal of Experimental Biology* 210 (2007): 3644–3660.

Rothman, C. M. "Live Sperm, Dead Bodies." *Journal of Andrology* 20, no. 4 (July–August 1999): 456–457.

Seymour, F. I. "Sterile Motile Spermatozoa Proved by Clinical Experimentation." *Journal of the American Medical Association* 112, no. 18 (1939): 1817–1819.

Sims, J. M. *Clinical Notes on Uterine Surgery with Special Reference to the Management of the Sterile Condition*. New York: William Wood & Co., 1867.

Spar, D. L. *The Baby Business: How Money, Science, and Politics Drive the Commerce of Conception*. Boston: Harvard Business School Press, 2006.

Stevenson, L. G., and R. P. Multhauf, eds. *Medicine Science and Culture: Historical Essays in Honor of Owsei Temkin*. Baltimore: Johns Hopkins University Press, 1986.

Stryker, J. "Regulations or Free Markets? An Uncomfortale Question for Sperm Banks." *Science Progress* (November 7, 2007).

Temkin, O., trans. *Soranus' Gynecology*. Baltimore: Johns Hopkins University Press, 1956.

Zachary, P. "A Most Unlikely Industry Finds It Can't Resist Globalization's Call—Exporting Human Sperm Is a Fast-Growth Business, Banks in Denmark, U.S. Find." *Wall Street Journal*, 6 January 2000, 1.

Chapter Fourteen: The Big Chill

Backus, L., L. A. Kondapalli, R. J. Chang, C. Coutifaris, R. Kazer, and T. K. Woodruff. "Oncofertility Consortium Concensus Statement: Guidelines for Ovarian Tissue Cryopreservation." *Cancer Treatment and Research* 138 (2007): 235–239.

Birrittieri, C. *What Every Woman Should Know about Her Fertility and Her Biological Clock.* Franklin Lakes, NJ: The Career Press, 2005.

Chen, C. "Pregnancy after Human Oocyte Cryopreservation." *Lancet* 1, no. 8486 (April 19, 1986): 884–886.

Gook, D. A. "Human Oocyte Cryopreservation." *Human Reproduction Update* 13, no. 6 (2007): 591–605.

Mundy, L. *Everything Conceivable: How Assisted Reproduction Is Changing Men, Women, and the World.* New York: Alfred A. Knopf, 2007.

Nieman, C. L., K. E. Kinahan , S. E. Yount, S. K. Rosenbloom, K. J. Yost, E. A. Hahn, T. Volpe, K. J. Dilley, L. Zoloth, and T. K. Woodruff. "Fertility Preservation and Adolescent Cancer Patients: Lessons from Adult Survivors of Childhood Cancer and Their Patients." *Cancer Treatment and Research* 138 (2007): 201–217.

Patrizio, P. "Molecular Methods for Selection of the Ideal Oocyte." *Reproductive Biomedicine Online* 15, no. 3 (2007): 346–353.

Porcu, E., and S. Venturoli. "Progress in Oocyte Cryopreservation." *Current Opinions in Obstetrics and Gynecology* 18 (2006): 273–279.

Porcu, E., et al. "Birth of a Healthy Female after Intracytoplasmic Sperm Injection of Cryopreserved Human Oocytes." *Fertility and Sterility* 68, no. 4 (October 1997): 725–726.

Sathananthan, A. H. "Ultrastructure of the Human Egg." *Human Cell* 10, no. 1 (1997): 21–38.

Sathananthan, A. H., et al. "From Oogonia to Mature Oocytes: Inactivation of the Maternal Centrosome in Humans." *Microscopy Research and Technique* 69 (2006): 396–407.

Spar, D. L. *The Baby Business: How Money, Science, and Politics Drive the Commerce of Conception.* Boston: Harvard Business School Press, 2006.

Wang, S. S. "Fertility Therapies under the Microscope." *Wall Street Journal,* 15 November 2007, 1.

Wells, D., and Pasquale Patrizio. "Gene Expression Profiling of Human Oocytes at Different Maturational Stages and after In Vitro Maturation." *American Journal of Obstetrics and Gynecology* 198, no. 4 (2008): 455.e1–455.e11.

Woodruff, T. "The Emergence of a New Interdiscipline: Oncofertility." *Cancer Treatment and Research* 138 (2007): 3–11.

Woodruff, T., and L. D. Shea. "The Role of the Extracellular Matrix in Ovarian Follicle Development." *Reproductive Science* 14 (2007) (8 Supplement): 6–10.

Xu, M., P. K. Kreeger, L. D. Shea, and T. K. Woodruff. "Tissue-Engineered Follicles Produce Live, Fertile Offspring." *Tissue Engineering* 10 (2007): 2739–2746.

Xu, M., T. Woodruff, and L. D. Shea. "Bioengineering and the Ovarian Follicle." *Cancer Treatment and Research* 138 (2007): 75–82.